# THE SEARCH FOR THE FIRST AMERICANS

# The Search for the
# First Americans

*Science, Power, Politics*

Robert V. Davis Jr.

UNIVERSITY OF OKLAHOMA PRESS : NORMAN

Library of Congress Cataloging-in-Publication Data

Names: Davis, Robert V., 1947– author.

Title: The search for the first Americans : science, power and politics /
    Robert V. Davis, Jr..

Description: Norman : University of Oklahoma Press, [2021] | Includes
    bibliographical references. | Summary: "Case study of the practice of science in
    its search for the First Americans and examines: (1) the conflicts between the
    methods of science and the traditional beliefs of modern Native Americans; (2)
    the power struggles for primacy of place internal to the sciences themselves; and
    (3) the interactions with external authorities such as government agencies, the
    press, universities, and museums. It examines how First American issues have
    been defined and how differences in cultural myths, scientific theories, research
    methodologies and public policy remain unsettled in modern America. It also
    investigates the blurred boundaries between science and myth as well as between
    fact and theory that ultimately weaken the credibility of science as a cultural
    mechanism for interpreting the natural world"— Provided by publisher.

Identifiers: LCCN 2021023226 (print) | LCCN 2021023227 (ebook) |
    ISBN 9780806175911 (hardcover) | ISBN 9780806175935 (pdf)

Subjects: LCSH: Indians of North America—Origin. | Indians of North America—
    History. | Indians of North America—Government relations. | Indians of North
    America—Ethnic identity. | Indians of North America—Science. | Indians
    of South America—Origin. | BISAC: HISTORY / Indigenous Peoples of the
    Americas | HISTORY / Civilization

Classification: LCC E61 .D25 2021 (print) | LCC E61 (ebook) |
    DDC 970.004/97—dc23

LC record available at https://lccn.loc.gov/2021023226

LC ebook record available at https://lccn.loc.gov/2021023227

# CONTENTS

Like many Americans, I have routinely followed popular press reports about early human remains found in the United States. As a young boy growing up in Oklahoma, I was fascinated by searching for arrowheads during camping trips and even enjoyed an occasional visit in Tulsa to the Thomas Gilcrease Institute of American History and Art, which houses one of the most extensive collections of American Indian artwork and artifacts in the world. I studied the symbolism in the Oklahoma state flag and the stories of the five "civilized" tribes. I uncritically accepted what I learned as the truth. However, as adults, even among people of goodwill, an agreed-upon truth can be elusive and may well be dependent on where and how we situate our conceptual lives. As Mark Twain reportedly warned, what gets us into trouble is not what we don't know, but rather what we think we know that just isn't so. The simple fascinations of our youth can lead later to complex intellectual journeys that disabuse us of some things that we thought we knew for sure. For me, this book has been just such an undertaking. In many ways it is a manifestation of the contradictions in American society between accepting the practical usefulness of science when it is in conflict with important cultural values.

The scientific search for the First Americans is, at a conceptual level, simple. As James Adovasio asked, "who the hell are these people, where did they come from, and when did they get here?" Although much has been documented, science is no closer today to finding an answer than at any time since the arrival of Columbus. In short, it has been a failed endeavor. Archaeology and anthropology, which are the intellectual domains most central to the search for the First Americans have, unfortunately, been caught fiercely protecting conclusions that are now known to be erroneous. Science has many strengths and is a critical component of the material progress that humans have enjoyed. But, at times, it can treat an explanatory theory with the same concreteness that should be reserved only for a material fact.

American Indians were present in the Western Hemisphere thousands of years before Columbus, or the Vikings, or the Phoenicians, or the ancient Chinese, or whomever has occasionally been put forward as the first to colonize the

New World. These indigenous tribes each have a creation story as to how they came to be. They do not tell stories of having wandered aimlessly over the Bering Strait during the last ice age. The difficulty for science is, understandably, valuing these beliefs when they are inherently contradictory and also conflict with clearly documentable findings of genetics and physics. Moreover, modern American Indians believe they are the direct descendants of the very first Americans and seem unwilling to concede that their ancestors might be later arrivals. This is clearly in conflict with the modern US psyche, which believes it must surely have been the destiny of everyone who ever came to this land to want to become an American.

The conflicting positions taken by First American scientists and modern American Indians have created an interstitial ontological void that has resulted in the power of the US political system being used to determine primacy of intellectual place. This examination of the search for the First Americans is a case study in the complex relationships between the biological findings of science and the social importance of cultural beliefs; the often fractious relationships across the sciences; the power of the US political process to dictate constraints on how science is conducted as well as how ancient human remains are to be disposed; and the hegemonic and often racist assumptions that are frequently embedded in how science has been conducted.

I must acknowledge the contributions of several others. I would like to thank the anonymous reviewers for their comments; Skip Furhman, Peter Schmitthenner, Ann Laberge, and Matthew Goodrum for their guidance on an earlier version; and Alessandra Tamulevich and Ihsan Taylor for their guidance through the publishing process.

While it is difficult, if not impossible, to operate completely external to one's system of values, it is my hope that I have fairly captured both the strengths and weaknesses represented in the positions taken by American Indians as well as First American scientists.

# Introduction

S
UBSTANTIVE DISAGREEMENT OVER A cultural past is a reflection of unstabilized power relationships in the present. This is a case study of the practice of science in this unstable environment as it has been employed in the search for the First Americans. It is not an attempt to determine the validity of any specific theory or material artifact; rather, it is an investigation into the practice of science as it relates to a search to explain a particular human existence. It is concerned with how personal, organizational, and cultural power has been manifested through scientific concepts, government policy, American Indian tribes, the public, and the scientists themselves in the search for the First Americans. It is an examination into how the issues have been constructed and, finally, the manner in which (as well as the degree to which) differences in cultural myths, scientific theories, research methodologies, and public policy remain unsettled in modern America. As this examination of the search for the First Americans shows, perceptions of the past can be as unstable as any projections of the future. Compounding the challenges faced by scientists, modern American Indians have used the US political system to gain a voice in defining the history of America's indigenous peoples. As a direct result, in 1990 the US Congress passed the Native American Graves Protection and Repatriation Act (NAGPRA) requiring that American Indian folklore and oral traditions be granted equal epistemological status with science in regard to the disposition of American Indian funereal artifacts and human remains found on federal and American Indian lands.

There is an inverse, and perhaps even a perverse, symmetry in the relationships created in the search for the First Americans. As anthropologist Randall McGuire has termed it, "the descendants of the conquerors study the ancestors of the conquered."[1] Moreover, the broader scientific community within which First American science must function is increasingly perceived as suspect. In President Obama's 2008 inaugural address, he stated, "We'll restore science to its rightful place, and wield technology's wonders."[2] What is science's rightful place in American society, and, specifically in the search for the First Americans, how was it dislodged from there? Can the federal government, on the one hand, "restore" science to some privileged position, but also enforce the provisions of NAGPRA on the other? One of the more challenging aspects is tracing instances of not only where observations from different scientific domains have led to validation or invalidation of a particular theory, but alternatively where observations in one domain have been shaped by the theories of a different domain.

The analytical tools associated with science studies especially lend themselves to deconstructing the practice of science in the search for the First Americans, the conflicts between science and the traditional beliefs of modern American Indians, and the power relationships internal to and among the sciences, American Indian groups, and public institutions such as federal and state governments, the press, universities, and museums. Although there are books on the search for the First Americans that focus on archaeology or are written for the popular press, there are no book-length treatments of this subject that address the dynamics across the scientific, cultural, and governmental realms; moreover, there are few offerings of even article length that approach the issues with the interdisciplinary tools of science studies. Among the few book-length analogues of the approach being taken are those of Naomi Oreskes in her examination of the twentieth-century geological community's fifty-year rejection of tectonic plate theory and Mario Biagioli in his placing of Galileo in the social context of the turn of the seventeenth century.[3]

Much of the necessary materials are the written products not only of American archaeologists, bioanthropologists, linguists, and geneticists, but also of American Indian scholars and the policy statements issued by modern American Indian organizations, the US Congress, the Department of the Interior, various US courts, and the popular press. Research dealing with the quest for the First Americans must address the myths of centuries, several sciences, the policies of the federal government, the politics of American Indian organizations, and the popular press, all within the context of the substantive science studies literature.

## Terminology

Given the intensity of the debate surrounding the issues, particular attention must be paid to the often unarticulated power relationships embedded in the terminology employed, and the extent, if any, to which there is a discernable influence on the practice of science. As philosopher Paul Feyerabend observed, subtle classification that is sensed rather than comprehended can affect perceptions.[4] Several issues associated with nomenclature have moved well beyond being considered subtle to being openly contentious.

A fundamental issue arises immediately around simply what to call the first people to enter the New World. Until the early 1960s, the name commonly used in the scientific community was Early Man. Today, anyone addressing this subject inevitably must choose between either Paleoamerican or Paleoindian to describe the first inhabitants of the Americas. All parties agree that such people are indeed Paleoamerican—that is, ancient American. However, some—especially those sympathetic to modern American Indian interests—consider the use of Paleoamerican to be a de facto statement that modern American Indians are not the direct descendants of the First Americans. Modern American Indians consider the proper term to be Paleoindian—ancient Indian—a recognition that their habitation and culture should be considered as unbroken in the Americas since the first humans arrived. Conversely, the defenders of using Paleoamerican believe that there is as yet no conclusive scientific basis for that position; modern American Indians could well be the descendants of a later, but still ancient, wave of migration from Asia and the very first American migrants may well have died out and left no modern descendants. Since many modern American Indian oral traditions hold that their cultural groups have been here continuously from the first peopling of the New World (whatever their version of "first" might be), *prima facie* acceptance of these oral traditions would lead one to conclude that Paleoindian and Paleoamerican capture identical concepts. However, some interpret usage of Paleoamerican as a political statement that modern American Indians are "old" Americans, but definitively not the oldest.

The issue became especially politically charged with the 1996 discovery of the remains of the Kennewick Man (as termed by scientists) or the Ancient One (according to American Indians) when they engaged in a legal confrontation over scientific access to the remains. Ironically, the terms Paleoindian and Paleoamerican, the distinctions between which are vitally important to American Indians, are both rooted in the very European-centric culture they disavow. Paleoamerican and Paleoindian are both designations by Western culture that create a chronological distinction between the people of an ancient past and ones of the modern present—a discontinuity that would appear to invalidate the very cultural continuity that American Indians espouse. Use of Paleoindian, the characterization preferred by American Indians, implies the existence of a more recent, and culturally distinct, Neoindian.[5] The nomenclature of First American science would appear to have the power to make the cultural past discontinuous with the present. Recognizing that there is no consensus, I will employ the term First Americans to describe the first humans to set foot in the Western Hemisphere.

Such a designation is sufficiently broad to accommodate, without prejudice, any descendants these First Americans may, or may not, have begat. The designation First Americans also attempts to capture the modern American sense that it was, of course, the destiny of the first inhabitants to become Americans.

A similar issue arises in the terminology identifying the modern indigenous peoples living in the United States. Depending on the audience being addressed, the names include Native Americans, First Peoples, First Nations, American Indians, Original People, Indigenous People, Aboriginal People, Amerindians, and Amerinds. Such putatively racist terms as savage and redskin have been widely discredited, even among sports teams, and are not seen in recent academic literature. Looking to the US governmental process for a politically adjudicated name is not particularly helpful. The name of the most recent substantively relevant legislation is Native American Graves and Repatriation and Act. However, the title of the relevant Smithsonian Institution museum is the National Museum of the American Indian. Moreover, important political activists and academic figures who claim an indigenous heritage, such as the late Vine Deloria Jr.—a member of the Sioux tribe; a prolific author; an academic scholar with degrees in science, theology and law; and a continuing challenger to Euro-centric thinking in the search for the First Americans—refer to their culture as Indian. I will follow the example of the Smithsonian Institution and Vine Deloria Jr. and use the term American Indian to describe the indigenous culture extant at the time of European arrival.

Contention also appears in constructing the terminology for the cultural periods. Perhaps the best example is the so-called Clovis culture, the long-reigning archaeological type-site in New Mexico that First American scientists for many years believed, incorrectly as it turned out, represented the cultural tool kit of the First Americans. The authority of this Clovis-first typology made it the standard against which all other finds were to be compared and interpreted. As a consequence, the Clovis-first paradigm produced accusations of anthropological imperialism on the part of US scholars. "Much of the discussion of the peopling of the New World uses North American terms (the label 'pre-Clovis,' generally applied, is irrelevant in South America where Clovis has never been found)."[6] Even the archaeological site of Monte Verde (ironically located in South America), which has now displaced Clovis as the earliest consensus First American archaeological site, is still referred to as a pre-Clovis site. For seventy years, the cultural authority granted to the Clovis-first paradigm so dominated the discussion that a sanctified chronological marker such as BC might well be borrowed and used by archaeologists to stand for Before Clovis. Nevertheless, although pre-Clovis has an inherent bias, the term is so prevalent that it will be used here.

One of the occasional difficulties in a research project is identifying a specific word that can express a key concept being brought to the foreground. A fundamental part of this project involves delineating the multifaceted relationships across myth, science, and other forms of cultural authority. While some may believe that myth has a taint about it of being a fable or a fairy tale—something important, perhaps, but also something not rooted in reality—here myth will be used to describe a statement concerning the natural world that is widely held within a culture and is generally accorded a status of being true, but at the same time is not necessarily (but might be) demonstrable according to the scientific methods of its time.

There is a continuing debate within the academic science studies community as to whether the range of methodologies that modern Western culture uses to investigate nature has such a commonality that it may be termed a single science, or whether it is sufficiently diverse that the proper term, therefore, is the plural sciences. In this distinction, the emphasis is on identifying a specific intellectual domain (e.g., physics, biology, chemistry) that is believed to characterize the natural world with a domain-specific set of methodologies. However, I am not examining the practice of science(s) through these domains, but rather through attempts by scientists, of whatever ilk, to solve one particular puzzle—the discovery and characterization of the First American. Any particular technique or finding may be claimed by physicists, archeologists, anthropologists, or biologists, and, in practice it can be, and frequently is, claimed by more than one discipline. To emphasize that the context of this examination of the practice of science is the puzzle of searching for the First Americans, and also to avoid being drawn into the conundrum of science versus the sciences, I will use the phrase First American science to capture the suite of methodologies employed. However, I will also examine the conflicts internal to First American science, which, at times, fracture along the traditional scientific disciplines espoused by individual scientists.

## First American Theories, Myths, and Evidence

Between the Introduction and the Conclusions in this book, there are three major sections. The first, chapters 1 through 5, examines and contrasts the various First American scientific theories against the American Indian creation myths, and assesses the evidence for each.

For people who are a product of the scientifically based Western culture, it is difficult to find common referents between science and American Indian creation myths involving entities such as Tiny Little Duck or the Spider Grandmother. Successful employment of truth statements is a powerful social lever. If a culture is structured around a particular form of truth, some element of that culture typically is empowered to translate truth in ways that dominate cultural values and, therefore, social behavior. The truth statements for both the practice of First American science and for the traditions of modern American Indians fall into this category. Some scholars contend that there is no inherent conflict between the bodies of knowledge and methodologies of science compared with that of creation myths because they are concerned with different matters.[7] As a practical matter, however, it is difficult to disentangle how these truths are translated into social power. As socially motive forces, the power-elite for both science and myth attempt to legitimize the basis of their truth statements by appealing to factors external to the statements themselves. For example, science appeals to the rigor of its methodology while cultural identity appeals to the sacred nature of its texts or oral traditions. Juxtaposing cultural authorities of such incongruent expressions of social power gives rise to contradiction and, therefore, epistemological conflict.

Central to this examination are archaeological sites at Clovis (Blackwater Draw), New Mexico; Monte Verde, Chile; and Meadowcroft Rockshelter, Pennsylvania, and how material evidence has been used, and, perhaps, abused by First American scientists to support conflicting theories. In addition, the status of Atlantis as a source for the First Americans is examined as both a scientific as well as a mythological statement. It is not that science is necessarily less speculative than myth, but rather, as Richard Feynman reportedly characterized it, science is imagination in a straitjacket. A review of the role of Atlantis will show that the methodological constraints imposed by the straight jacket change through time, with substantive implications.

The modern self-image of service to humanity that is embedded in some scientists was already evident in the nineteenth century. On October 21, 1898, aboard the steamer *Germanic*, Jeremiah Curtin dedicated a book titled *Creation Myths of Primitive America* to John Wesley Powell, who had lost an arm serving in the Union forces in the American Civil War and subsequently directed the US Geological Survey as well as the Bureau of Ethnology at the Smithsonian Institution. In the dedication, Curtin wrote: "You lost your right hand in fighting to save American unity.... The same kind of impulse that sent you to the field of battle to serve the country and world, sent you to the field of science to serve as a geologist and an explorer. . . . Through your labors, combined with those of the men whom you have associated with you, the world has learned more of the great primitive race of our country than it learned from the discovery of the continent till the day when the Bureau was founded."[8] Curtin's dedication revealed two important points. First, he demonstrated a recognition of the influence of the scientific elite of the community through his sycophantic appeal to the good graces of Powell, certainly one of the dominant figures in American geology, archaeology, and ethnography of that period. The second was a declaration of the self-mindful role of then-modern science in providing the definitive statement on the "primitive race" in America; there is no mention of American Indian myth as an epistemological resource even though the book, which includes Curtin's dedication, is about the creation myths of these very same people.

## First American Science

Chapters 6 through 9 examine the unstabilized power relationships between the several sciences as they are employed in the search for the First Americans. Although the written reports of First American scientists reflect a widespread representation of scientific production, they are not the only sources to be explored. First American science, especially the field of archaeology, at its core is a science of materiality, of artifacts. Moreover, it is not only the artifacts themselves that are important in the scientific debate, but also their physical context—their in situ relationships. One manifestation of this physicality is the archaeological sites where not only the artifacts themselves are recovered, but also where the laboratory techniques of recovery and the positional relationships are available for inspection by the First American scientific community. A second manifestation of this physicality is the prominent display of First American-related artifacts at publicly accessible museums. As a representative sample of this physical aspect of First American science, I visited what is regarded as the Clovis-culture type-site at Blackwater Draw, New Mexico, and also what is now considered to be a significant pre-Clovis site at Meadowcroft Rockshelter, Pennsylvania. In addition, I visited the Hall of Human Origins at the Museum of Natural History and the National Museum of the American Indian, each a part of the Smithsonian Institution in Washington, DC, and I also visited the Blackwater Draw Museum in Blackwater Draw, New Mexico.

One of Michel Foucault's contributions to the history of science is the concept of an epistemological grid in Western societies.[9] According to Foucault, there are unquestioned and powerful normative constructions that are so fundamental to a cultural fabric that a concerted effort is required to separate them from the societal background noise. Unless the underlying values are identified, historians have little prospect of differentiating meaningfully between the characteristics of modern science and the broader contextual mores of society at large. Although scholars have attempted to create a distinction between what is external and what is internal to science, this distinction has largely been discredited in science studies. This raises the issue as to whether the very concept of a *first* American is value neutral. Does the problem itself have a set of contextual assumptions that structures whether, and what kind of, science can answer it (or not), what constitutes an observation, and what might comprise an acceptable answer? In some quarters, it is inherently unacceptable, even racist, to conduct research examining whether the First Americans might have migrated from what is now considered to be Europe, or to conclude that the First Americans migrated into the New World from Asia across a Bering Strait land bridge.

## Community

Chapters 10 through 12 examine the concept of community, which is of central importance to many aspects of First American science. It is an important social mechanism that not only shapes the boundaries of science and myth, but also validates who is considered to be a scientist or a member of a tribe. A community motivates and rewards, and it produces a value system that legitimizes or rejects theories, myths, observations, and methods. Any objectivity that First American science is accorded is a function of a group's practice of science rather than of any single individual's efforts.[10] A community has power-structures, institutions, elites, and normative constraints. Understanding community dynamics and their constituent elements is an important part of understanding science in the context of its search for the First Americans.

With a near consensus in the science studies literature that community values play some role in the practice of modern Western science, it is important to examine how these values are, if at all, modified from time to time and how fluid might be the modification process. In the United States, it is clear that the government—primarily, but not exclusively, at the federal level—is significantly involved not only in funding scientific projects and shaping their goals, but also in creating acceptance for scientific methodologies. Science may be considered as speaking truth to the power of the federal government, but that same political power has great influence over how First American science is practiced.

In the United States, federal law is the most intrusive form for externally dictating cognitive values to the scientific community. US law grants American Indian traditions and oral history equal epistemological status with the methods of First American science for determining the cultural affiliation of, and therefore access to, pre-European human remains and funerary objects found on federal or American Indian lands. There have, of course, been objections to governmental intervention from members of the scientific community, as captured by the comment of Douglas W. Owsley of the National Museum of Natural History, Smithsonian Institution: "Positions taken by those persons or organizations opposing careful scientific study undermine attempts to learn about the pre-history and settlement of this hemisphere."[11] Taking a diametrically opposing position in *Red Earth, White Lies*, Deloria criticized the demeaning role of science and trumpeted the ontological value of American Indian traditions.[12] This, in turn, led James Adovasio, a First American scientist who is excavating the potentially pre-Clovis Meadowcroft Rockshelter, to conclude that Deloria was engaged in "an impassioned (and largely ill-informed) diatribe against the matter of human evolution, the existence of the Bering land bridge, and the continuing failure of the sciences to confirm the creation stories of Native American myth. . . . In response to such assaults, he denied the underpinning validity of any and all science, its findings, and its methods."[13] Obviously participating in the search for the First Americans is not for the socially sensitive.

## Bibliography

There is an abundance of literature on the science, myths, and politics of the search for the First Americans. However, it tends to be scattered across the several scientific disciplines, American Indian organizations, and the federal government—and covers a five-hundred-year period. I have included an extensive bibliography that addresses the entire range, both chronological and disciplinary, of materials available.

# First American Theories, Myths, and Evidence

# CHAPTER I

# American Indian Creation Myths

C ULTURES HOLD, AND EVEN REVERE, existential myths. It may be a source of power for a dominant class or religion, provide a comforting structure for social continuity, or reinforce a sense of cultural superiority. At the C. G. Jung Institute during the winter of 1961–62, Marie-Louise von Franz (1915–98), a noted Jungian psychologist and scholar, gave a series of lectures on creation myths that were published in 1972 as *Creation Myths*. According to von Franz, "creation myths are of a different class from other myths. . . . [T]here is always a certain solemnity that gives them a central importance. . . . [A]s far as the feeling and emotional mood which accompany them are concerned, creation myths are the deepest and most important of all myths."[1] Deloria, echoing the conclusions of von Franz, stated that "every human society maintains its sense of identity with a set of stories which explain, at least with satisfaction, how things came to be. Many societies begin at a creation and carry forward a tenuous link of events which they considered to be historical."[2] Yet, in the United States there is a particular fascination surrounding the First Americans that is difficult to understand in these terms. This intense interest by a dominant European-centric culture is in identifying an exact source for the First Americans—the people of another culture.

At the times of the commonly acknowledged European arrivals, approximately 1000 CE by the Vikings and 1492 CE by Christopher Columbus, there was no European presence—linguistically, culturally, or phenotypically. For the conquering Western European culture, the roots of the Judeo-Christian imperative to dominate nature ran deep.[3] As Dana Villa explained, the European conquest of the Western Hemisphere objectified the earth itself.[4] It represented perhaps the last great barrier to human dominion over its physical circumstance. It may be difficult, however, for a culture to distinguish between its self-perceived superiority and the grandeur of Homo sapiens at large. There is no civilization in which its creation myth recounts that the god(s) passed them by and chose, instead, to anoint a neighboring civilization. The cultural authority that sustains an existential myth does not provide for a people to have been an inconsequentially distant relation of the first people; there must be a direct lineal descent from the creative primogenitor. As a cultural act, the social construction of a society's creation myth is one of the few defining events in its history. However, when that myth is successfully challenged, then the creation myths of other societies are suspect as well. As pointed out by Deloria, "when secular science defeated Christian fundamentalism, in its victory it was able to promulgate the belief that all accounts of the creation or of spectacular catastrophic specific events were superstitions devised by ignorant peoples to explain the processes of the world around."[5]

The characteristics of creation myths, according to von Franz, are common across cultures. Typically, they fall into such categories as the awakening of consciousness; creation from above; creation from below; germs and eggs; or a mechanistic god, such as the Chinese P'an Ku, who is a master in all crafts, and the Egyptian God Ptah, who used a potter's wheel to create the world and the other gods.[6] American Indian creation myths generally conform to this model and can be categorized as describing an existential birth by: jumping, falling, or being pushed from the sky; arising from the earth or mud; appearing from the lake, ocean, or river; or being born in and climbing out of an underworld.[7] Simultaneous with this creation, in many American Indian traditions they are also given an immediate awareness that they have been created.[8] "Some have a creator, and some don't and a few seem to have grafted a creator onto an older story that had none, perhaps under the influence of Christianity."[9] There also appears to be no relationship between the geographical area or linguistic roots, on the one hand, and the particular category of creation myth indigenous to an American Indian tribe, on the other.[10]

Jeremiah Curtin's 1898 *Creation Myths* is a straightforward and readable account of twenty-two creation stories of American Indian groups. Considering the obvious racial and other social biases typical of his time, outside of the "Introduction" chapter, there is no overt attempt to provide a pejorative interpretation of any of these myths; the only seeming constraint would be that faced by all ethnologists—Curtin's ability to accurately capture the cultural values and priorities of non-Western civilizations within the conceptual boundaries imposed by an American-culture and English-language epistemological grid. Compounding this difficulty is the contention by American Indian scholars that as much as 90 percent of traditional American Indian "information" has never been printed and is therefore not available to scientists or the public.[11] To the extent that this is the case, it would further constrain the ability of Western scholars to understand American Indian myths.

The existence of Curtin's *Creation Myths* resulted from the continuing popular interest in the indigenous cultures of America. Based upon interviews of "Indians" in California, Mexico, and Guatemala, it is a compilation of consecutive Sunday articles written in 1895 for the *New York Sun* of "myth-tales as I might think of sufficient value to appear in the paper."[12] Curtin employed a surprisingly modern concept of what he termed the American *system* to capture the intricate set of relationships between the mental and physical domains that are necessary to create a complex myth. According to Curtin, "The American creation myths, as far as we know them, form simply a series of accounts of the conflicts, happenings, and various methods by which the first world was changed into the world now existing."[13] Embedded within these creation myths, Curtin believed, were "models upon which faithful Indians are to fashion their lives at all times and places."[14] These myths were not simply interesting stories, but rather formed an essential element of the American Indian value system.

The descriptive range of creative agency for American Indians is quite extensive. For the Pawnee, it was Tiny Little Duck; for the Crow, it was Old Man Coyote; Mandan, the Lone Man; Yakima, the Wooden Stick People; and Hopi, the Spider Grandmother.[15] Knud Rasmussen, in his book *The Eagle's Gift*, asked a Noatak River Eskimo named Apatac how the world was created. "Nobody can know anything for sure about the beginning of life. But whoever opens his eyes and his ears and tries to remember what the old people said, might fill the emptiness of his thought by this or that knowledge."[16] For the American Indian, Curtin concluded that "The wisdom of his nation is more valid, more reliable than the wiriness of his own senses. His eyes and ears might be deceived by tricksters, but not by the truth delivered to great men among his own people, preserved by them sacredly and passed down to others."[17]

Why do Americans study American Indian myths? As Curtin captured it in 1898, at the time that the Europeans arrived, the myth systems of the American Indians were intact. There was no culturally sanctioned process by which they were challenged. "Human history has no second example of a single system of thought developed over such a vast area."[18] For modern Americans, the appeal of the search for the First Americans is still rooted in a quest for confirmation of a cultural existential myth.

The late University of Cambridge physicist Stephen Hawking began *A Brief History of Time* with the story of a skeptic who challenged a philosopher (Hawking hypothesized that it was Bertrand Russell) with the statement that the "world is really a flat plate supported on the back of a giant tortoise." When asked what supported the tortoise, the skeptic responded that "it's turtles all the way down!"[19] The existential myths of American Indians suffer from the same potential epistemological challenge. Although their traditions identify a primal creative agency, the same myths do not typically explain the origin of the usually animate agency, nor do they explain the occasional presence of other contemporaneous aboriginal tribes in their myths. It is, perhaps, this inability of American Indians to provide an explanation that relates their creation stories to their demonstrable material surroundings that provides First American science its most significant lever for challenging these myths.

There is discipline as well as power in capturing concepts through the written word. Bruno Latour, Steve Woolgar, and others have described science as ultimately the sum of its written reports, its so-called inscription devices.[20] The act of writing denaturalizes a myth.[21] Commitment to writing is one form of assigning boundaries to the physical possibilities of an oral tradition. Scientific analysis is another and is, perhaps, even more constraining than being captured by words. For a Western, science-based culture, the contest is perceived as being between laboratories and written reports, on the one hand, and campfires and oral traditions on the other.

The power to define the past is of value as a mechanism for negotiating current power relationships and, therefore, influencing the future. For American Indians, myths are a part of an enduring social fabric and are nonargumentative. They have a status that is believed to be a nondiscoverable part of human knowledge. This does not, however, mean that there is no social value to myths. Even in a science-based culture, they may be part of an ethical structure that governs the mores of that society. According to Gerard Naddaf, a professor of philosophy at York University, myth can provide both a causal explanation for the present social, and even natural, order along with some assurance of legitimacy and continuity. Myth operates to integrate potentially conflicting elements of the religious, cultural, historical, social, and political realms.[22] The challenge that First American science presents to American Indian myths is that the material artifacts analyzed by modern science may destabilize a long-standing social order with vested interests. For some cultures, principally ancient but some modern as well, something accorded the status of myth is not irrational, but rather is considered to be a fundamental part of reality, "a liberation from excessive abstraction and objectivism," according to Kathryn Morgan, a professor of classics at UCLA.[23] For such cultures, myth creation and maintenance is an ontologically definitive process.

The Smithsonian Institution National Museum of the American Indian (NMAI) is a representational gallery of American Indian culture more than it is a traditional science-based museum. The exhibits were prepared by community curators approved by, and often supplied by, the tribe whose culture was being exhibited. As such, it provides an interesting indigenous insight into some of the specific American Indian creation myths. A placard by the Cherokee tribe described the roles that Water Beetle and Buzzard played in aiding "the bird and animal people," presumably including the Cherokee, to descend from above the sky and inhabit the world.[24] At another Smithsonian Institution NMAI exhibit, the Tohono O'odham tribe provided a water-borne depiction of human origins through the creational agencies of Earth Medicine Man, Buzzard, Coyote, and termites.[25] In yet two more exhibits, the Ka'apor tribe of Brazil portrayed its own creation as being "from the beginning of time," with the creative agency provided by "Mair,"[26] and the Pamunkey tribe of Virginia offered a less mythic statement as to its presence in what is now Virginia for "at least 12,000 years."[27] It was not clear as to whether the Pamunkey historical memory has an actual calculation that reaches back to that time, or whether they are basing it on the results of First American science. Regardless, such an antiquity would chronologically place their culture as one of the original inhabitants of the New World. Finally, in what might be considered a bit of hyperbole, the Smithsonian Institution NMAI itself stated that "the true significance of 1492" is its being "the most profound event in human history."[28] Although other events (e.g., the migration of Homo sapiens itself out of Africa) could arguably be considered of more import, the statement itself provides an indication of the cultural importance that modern American culture, as expressed through its preeminent museum system, attaches to the search for the First Americans.

# CHAPTER 2

# Euro-American Theories

THE CHALLENGE FOR FIRST American scientists is inelegantly, but aptly, captured by James Adovasio, who asked, "Who the hell are these people, where did they come from, and when did they get here?"[1] This chapter will lay out how the answers to these key questions in the search for the First Americans have been continually revised within the dominant Euro-American culture beginning at the time of the initial fifteenth-century contact and continuing even through today.[2]

One theoretical contention that need not be addressed here in detail is human evolution per se; there is a consensus, at least to date, that human skeletal remains and material artifacts found in the Americas are all associated with modern Homo sapiens.[3] As a result, the substantive challenges facing First American scientists are determinations of the propinquity of cultural relationships rather than determinations of corporeal evolution. The anatomical and genetic distinctions that receive so much attention in defining the Homo sapiens ancestral tree in Africa, Europe, and Asia play a much different role in the search for the First Americans. Physical distinctions among early Americans, while of some interest in and of themselves, are for First American scientists tools to be used to make estimations of social relationships. Consequently, much of the five-hundred-year-old debate within the Euro-American culture has been consumed with conforming what is believed to have been discovered about the First Americans to the foundational cultural truths of the dominant Euro-American culture. A popular adage attributed variously to both Mark Twain and Will Rogers states that what gets us into trouble is not what we don't know, but rather what we think we know for sure that just isn't so. The Euro-American cultural baseline—what is accepted as being known for sure—has been unstable in the search for the First Americans for the past five centuries as a result of the continuing redefinition of the cultural ontological judgments derived from biblical tenets, racial hierarchies, evolutionary theory, and the development of modern geological determinations of an ancient earth.

Edwin N. Wilmsen, in his 1965 article in *American Antiquity*, divided the five centuries of vacillating Euro-American thoughts on the First Americans into six periods: 1520–1780, 1780–1859, 1859–90, 1890–1925, 1925–50, and 1950 to the present.[4] The first phase, 1520–1780, centered on the conjectures of Europeans who, for the most part, had never been to the New World. As the Americas became populated by educated men, from 1780 to 1859 the theorizing began to be based on direct observations, such as those by Thomas Jefferson, associated with the nascent development of modern anthropology, archaeology, and geology. Wilmsen's next phase in the search for the First Americans, 1859–90, grappled with the implications of biological evolution and geological antiquity. Against this background, the search began for New World analogues to the ancient human remains being found in Europe. Educated, but professionally untrained, amateurs became involved in "discovery" of artifacts on which exaggerated, and routinely disproven, claims of antiquity were built. Between 1890 and 1925, the almost unconstrained speculation of the prior period led, according to Wilmsen, to the creation within the newly developing community of First American scientists of a proscription against the possibility of truly ancient (Pleistocene) human remains in the Americas. It was during this period that key figures such as William Henry Holmes (1846–1933) and Aleš Hrdlička (1869–1943), a trained physician, developed their intellectual approaches. The next period, 1925–50, not only brought professionalism to archaeology, but also laid the foundation for what has been an almost visceral skepticism by leading First American scientists for the last seventy-five years of the twentieth century toward any evidence that suggested human occupation of the New World prior to thirteen thousand years ago, and an autonomic disbelief in any theory that contradicted the thesis that the Clovis culture—so named because of its initial discovery near Clovis, New Mexico—was the culture of the First Americans.[5]

The primary contention for the proposed route of the First Americans into the New World, as conjectured by many Euro-American scientists, has been that of a footpath across what is now the Bering Strait on a land bridge that was exposed during the last ice age, followed by a rapid expansion southward between the Cordilleran Glacier and the Laurentide Ice Sheet. However, there are also alternative theories for possible northern-latitude maritime entries from both Asia and Europe that bypass the Beringian land bridge.[6]

## Early Theories (1500 to 1800)

The concept of a "New" World, frequently attributed to Peter Martyr (1457–1526) in *De Orbe Novo* (*On the New World*), was a late fifteenth-century,

European-centric statement of cultural, temporal, and geographical hegemony over Western Hemisphere civilizations.[7] As it became apparent over the next few decades after the publication of this book that this large land mass was not contiguous with the Old World, explanations that accommodated European cultural myths were needed regarding the origins of the New World's indigenous peoples. Speculation included the Lost Tribes of Israel, Romans, Phoenicians, Mongols, Welsh, Hindus, Carthaginians, and survivors of the lost continents of Mu and Atlantis.[8] By 1578 William Bourne published in *London Booke Called the Treasure for Travellers* an entire chapter titled "The Peopling of America."[9] The first significant documented conjecture that the indigenous Americans were originally from Asia came in 1590 when, after living in what are now Mexico and Peru, Jesuit missionary José de Acosta (1539–1600) speculated that they had migrated across something like what is now known as the Bering Strait between Russia and Alaska.[10]

One of the difficulties faced was not only a determination of the geographical origins of the inhabitants of the Western Hemisphere, but also a determination of their genealogical origins. In short, regardless of their physical origins, a seminal question to be addressed was the extent to which the aboriginal peoples could be considered human. In the absence of any clear biblical reference, the prevailing belief for the initial Spanish settlers was that the American Indians were not human. Indeed, a few even supported an Aristotelian view of spontaneous generation from mud. Auroleus Phillipus Theostratus Bombastus von Hohenheim (1493–1541), known as Paracelsus, espoused the radical opinion that American Indians were not descendants of the biblical Adam and Eve, but rather had been separately created and had no souls.[11] According to Paracelsus, "It is most probable that they are descended from another Adam. . . . [I]t is credible that they were born there after the deluge, and perhaps they have no souls."[12] Such a view facilitated treating the American Indians as an inferior species to be exploited as a part of the Spanish and Portuguese economic development of the New World. Some, such as the Dominican Bartolomé de las Casas (c. 1484–1566), opposed such a view and personally supported the Lost Tribes of Israel as the origin.[13] This, however, again presented the challenge of locating their more specific geographic and genealogical origins. The convenient solution of a nonhuman origin gradually was no longer considered viable.

During the 1640s, there was an intense debate in the Netherlands. Hugo Grotius posited that the American Indians were principally Scandinavians with Ethiopian, Chinese, and Moluccan influences. Johannes de Laet and George Horn countered that American Indian origins were a combination of

Scythians, pre-Columbian Spanish, Welsh, and Polynesians. In 1702, Cotton Mather (1663–1728) wrote in *Magnalia Christi Americana* that "probably the Devil decoyed these miserable savages hither, in hopes that the gospel of the Lord Jesus Christ would never come here to destroy or disturb his absolute empire over them."[14] Thomas Hobbes (1588–1679) reflected the embedded conception of "savage" peoples of his time in doubting that the indigenous inhabitants of the Americas had possessed the necessary cultural sophistication to build the ancient mounds and monuments of the Americas discovered by European explorers, leading him to conclude that their life was "solitary, poor, nasty, brutish, and short. . . . For the savage people in many places of America . . . have no government at all, and live at this day in that brutish manner, as I said before."[15] Georges-Louis Leclerc (1707–88), known as Comte de Buffon and one of the critical figures in the establishment of modern biology, wrote in his influential multivolume *Histoire naturelle, générale et particulière* that both the people and the animals in the New World were inferior to those of the Old.[16]

Despite this contentious assortment of First American origin possibilities, Asia as the primal source continued to figure prominently. In 1648, the Russian Semyon Dezhnyov (c. 1605–72) is now believed to have been the first European to have sailed through the narrow channel separating the Asian and North American landmasses. However, there apparently was little recognition at the time of his feat. In 1728, Vitus Bering (1681–1741) successfully traversed that same strait, which is now accorded his name. As knowledge of Bering's successful trip began to spread, the Asian connection found ready support and was echoed in speculation such as that of Thomas Jefferson in his 1784 *Notes on the State of Virginia*.[17] In 1794, Jesuit Ignaz Pfefferkorn (c. 1725–98) wrote in *Description of the Province of Sonora* that "to me it is almost certain that the first inhabitants of America really came by way of this strait."[18] But this view still remained only one suggestion among many.

## Theories of the Nineteenth Century

The nineteenth century was an especially dynamic period in the development of what has become First American science. Biology, physics, chemistry, anthropology, archaeology, philosophy, and geology—the constituent elements of First American science—were each stabilizing into an independent discipline utilizing a particular suite of methodologies to address specific puzzles presented by the physical world. During this period several important advances were made in these nascent disciplines. First, geologists such as Georges Cuvier (1769–1832) and Charles Lyell (1797–1875) were breaching Bishop James Ussher's (1581–1656) six-thousand-year barrier and establishing the age of the earth as being at least in the millions of years. Second, Charles Darwin (1809–82), Alfred Russel Wallace (1823–1913), and others employed this expansion of geological time as a critical element to support their theory of biological evolution. Finally, archaeologists such as Jacques Boucher de Perthes (1788–1868), Hugh Falconer (1808–65), William Pengelly (1812–94), and Joseph Prestwich (1812–96) were finding ancient tools that could be dated against the chronology of the newly defined geological record.[19]

Human origins history is replete with anecdotes that capture the emotions and prejudices of a particular time. Lyell (1797–1875) barred women and children, due to their sensitive nature, from his public lectures because of the sacrilegious implications of his findings.[20] According to Adovasio, "It is little wonder, then, that when white Americans came across the most monumental works of the original inhabitants of North America, they assumed them to be the product of some other, master race long since vanished: the mound builders."[21] Built upon a nineteenth-century framework of cultural beliefs that included a hierarchy of races, the conceptual options available to early First American scientists included only those that differentiated pejoratively between the indigenous inhabitants of the Western Hemisphere and the Caucasian creators of the dominant Euro-centric culture.

With the professionally trained scientist replacing the educated gentleman of leisure as the principal investigator of natural phenomena, with the ever-increasing evidence of the antiquity of humans in Africa, Europe, and Asia, and with the European assessment of the primitive nature of the indigenous cultures of the Western Hemisphere, there was little reason to doubt that evidence would be discovered confirming a comparable human antiquity in the New World. In North America, the most obvious manifestations of an ancient civilization were the burial mounds scattered throughout the continent. Prominent in the archaeological histories were the excavations of Caleb Atwater. Both Adovasio and Wilmsen cited Atwater's 1820 *Transaction from American Antiquarian Society in Boston*, later self-published in 1833 as *Writings of Caleb Atwater*.[22] Based on his examination of the mounds along the Ohio River, Atwater hypothesized that North America had been populated via the Bering Strait first by pastoral farmers and shepherds and then by savage hunters migrating from India who were the ancestors of the modern American Indian.[23] Wilmsen credited Atwater with providing "a new perspective" based on what "were probably the first systematically controlled archaeological observations in the Americas."[24]

As the source of the First Americans, a recurring myth has been that of a Jewish origin for the New World inhabitants. According to the American scholar and politician Ignatius Donnelly (1831–1901), there was "scarcely a custom known to the Jews that does not find its counterpart among the people of the New World."[25] However, Donnelly challenged any direct Jewish colonization of the Americas, not believing that "they left their flocks and herds, marched across the whole face of Asia, took ships and sailed across the greatest of the oceans to a continent of the existence of which they had no knowledge."[26] Donnelly then proceeded to attribute what he believed to be similarities between Jewish and American Indian cultures to the fact that both had a common cultural antecedent—Atlantis. His analysis was clearly intended to demonstrate cultural similarities between the two groups but supporting only an indirect colonization of the New World by Jews. Where Donnelly concluded that Jews populated the New World via Atlantis, the Book of Mormon described Jews as migrating directly across the ocean. The Mormon view of the Jewish migration is but a part of a more general apocalyptic war between God and Lucifer that saw a division of humanity into light-skinned and dark-skinned, including American Indian, peoples. The Book of Mormon told of Jews crossing the ocean to North America and then splitting into the godly Nephites and the savage Lamanites that were marked by a red skin. For many years Mormons reportedly told American Indians that acceptance into the Church of Jesus Christ of Latter-day Saints would lighten their skin.[27]

Along with the First American theories associated with the Lamanites, Jews, and Atlantis, there were many others offered by both reputable scientists as well as what might charitably be termed intellectual speculators. What can be judged as a scientifically acceptable view is a function of a particular time, place, and set of social values. For example, John Wells Foster (1815–73), an early geologist and paleontologist who in retrospect held what would now be considered highly racist views, catalogued many of the First American origin theories as being no more than conjecture. Foster dismissed notions that the First Americans had been Northmen arriving via Vinland or Welsh serving under Prince Madog in the southern United States; and he also rejected such purported evidence as ancient Danish runes on Dighton Rock in Massachusetts; Hebrew letters on the Holy Stone of Newark in Ohio; the Cincinnati tablet with Egyptian-like cartouches in Ohio; and the Round Tower of Newport built in Rhode Island by the Norse.[28]

With the proliferation of theories as to the origins of the First Americans and their relationship, if any, to modern American Indians, scholars began to catalogue the disparate narratives. One of the earliest comprehensive efforts was that of Justin Winsor (1831–1897), librarian of Harvard University, in the eight-volume *Narrative and Critical History of America* that was for many years a standard reference text for those studying American history.[29] Published between 1884 and 1889, Winsor, as both author and editor of the volumes, described not only the explorations and settlements of the European nations, but also specifically addressed the search for the First Americans in the 1889 Volume 1 entitled "Aboriginal America." In the evolution of First American science, the nineteenth century was an interesting mélange of unsupportable speculation occurring simultaneously with the development of the initial scientific methodologies on which the foundations of First American science would ultimately rest.

## The Twentieth Century

Archaeological discoveries in 1929 near Clovis, New Mexico, provided a scientifically supported theory that the first settlers were the "Clovis" people who had migrated thirteen thousand years ago from Asia across a Bering Strait land bridge at the end of the last ice age, and that they had swept through Alaska, populating all of the Americas in a mere one thousand years.[30] Despite what, in retrospect, was significant contradictory evidence, the Clovis-first paradigm bested all challenges for seventy-five years and is still supported by a dwindling number of First American scientists. While these debates were largely archaeologically centered, genetics, linguistics, and bioanthropology have increasingly generated both supporting and contradictory observations for virtually every theory.

As both scientific and popular First American theories have evolved since the late fifteenth century, the predominant, but not sole, narrative has been that of an Asian source and a lack of comparable antiquity compared with the peopling of the Old World. "The near-universal consensus that America's first peoples were bygone Asians occurs no matter the differences of evidence and opinion regarding exactly where in Asia they originated."[31] Even if such a consensus indeed existed, there would still remain the debate surrounding the physical path (land bridge across the Bering Strait or coastal maritime), the antiquity (pre-, mid-, or post-ice age transit), the number of migratory pulses from Asia, and the extent, if any, to which modern American Indians are the direct descendants of the First Americans. "Of the possible entry routes into the Americas, Beringia, a land bridge from Siberia to the interior and coastal areas of Alaska and northwest Canada, is the most viable. Entry from the Atlantic side of the hemisphere is a competing proposal."[32] Although constituting a distinctly minority opinion, reputable scientists have supported this alternative theory citing Europe as a source of the First Americans.

It is difficult to create a functional taxonomy of First American theories. Establishing taxonomic relationships requires identification of commonality across the entities being evaluated, even if the relationship is as simple as a binary presence or absence of some quality. Some First American scientists would seem not to have heard of the aphorism that the absence of evidence is not evidence of absence. A declaration of the *first* Americans is simultaneously a definition of *non*-Americans, a declaration that something does not exist, that no other earlier culture will be found. The difficulty in accepting such an approach is that in 1908 what was then believed to be evidence of the First Americans at Folsom, New Mexico, was replaced in 1929 by the First Americans at Clovis, New Mexico, who, in turn, have recently been replaced by the First Americans at Mont Verde, Chile, who might, someday soon, be supplanted by the First Americans at Meadowcroft Rockshelter in Pennsylvania, the First Americans at Buttermilk Creek in Texas, or most recently the First Americans at Chiquihuite Cave in Mexico

Although determination of chronological relationships is paramount in First American science, *quantitative* metrics that might be used for structuring the relationships are either missing or rely heavily on so-called expert opinion. Potentially differentiating, and frequently contradictory, *qualitative* elements of the theories that have been offered include migration patterns; existential myths; and colonial, Western, indigenous, religious, biblical, historical, scientific, and economic sentiments. Although there have been efforts to create a taxonomy of the physical structures of classes of artifacts, there is disagreement among the experts over even such a straightforward judgment as to what constitutes "fluting" (a narrowing of the base) on a stone spearhead. Furthermore, because of the politically charged atmosphere surrounding the search for the First Americans, judgments on a single physical factor such as fluting have also been perceived as a way to assert an American hegemonic umbrella over the identity of the first inhabitants throughout the entire New World.

With the increasing professionalization of First American science at the beginning of the twentieth century, the elite of the profession began to question human antiquity in the New World. In the late nineteenth century, William Henry Holmes began assuming ever more powerful positions under the tutelage of John Wesley Powell, first in the US Geological Survey and then in the Smithsonian Institution Bureau of American Ethnology, and indicated a clear disbelief that humans had been in the Americas as long as in Europe.[33] In 1903, Holmes appointed Aleš Hrdlička as assistant curator in charge of physical anthropology at the Smithsonian Institution National Museum. As a result of this official position, a forceful personality, and his unwavering theoretical position, for decades Hrdlička dominated both the doctrine and the methodologies associated with the search for the First Americans. Echoing the position taken by Holmes, Hrdlička aggressively and successfully opposed those who argued for the great antiquity of the First Americans. For Hrdlička, it was clear that analysis of the physiological, cultural, and linguistic characteristics of the indigenous Americans demonstrated an absence of the diversity that would result from great antiquity.[34] As a result, very early in the twentieth century, "Early Man in America was dead so far as most anthropologists were concerned."[35] To his credit, not only has no meaningful evidence, to date, been discovered to refute Hrdlička's skepticism concerning a great antiquity for the First Americans, he was also largely responsible for the professionalization and institutionalization of the methodologies of First American science. His efforts, however, have not been without criticism. As archaeologist James Adovasio characterized it, "One can, a bit wildly, think of Holmes and Hrdlička as the Doc Holliday and Wyatt Earp of early-twentieth-century anthropology, running all abusers of the proper scientific method out of Tombstone."[36] During the first half of the century, First American science was largely a top-down enterprise.

As the twentieth-century debates over human origins stabilized worldwide, for Africa the chronological range covered a few millions of years, and in Europe the debate centered over the past few hundred thousand years. But for the New World, the debate narrowed to only the most recent few tens of thousands of years. Despite this great temporal constriction for the arrival of the First Americans, the intensity of the acrimonious debate was not proportionately reduced. Writing in 1989, archaeologist David Meltzer stated that "Advocates of a pre-12,000 BP [before present] human population assert that their evidence is valid and is rejected by skeptics only because of deep-set historical biases. That assertion is not well-founded."[37] Adovasio in 2003 vehemently countered that the long-reigning theory that no pre-12,000 BP Americans existed had no basis in fact. "Such a belief is, of course, not science. And it is not logical. It is, in fact, more like a religious dogma."[38] Meltzer, however, believed the process of First American science to be much more deliberative than that described by Adovasio. For Meltzer, "Each new candidate for great antiquity brings with it fresh claims, but the outcome remains the same. Skeptics raise questions. Debate ensues. The claim is accepted by some, rejected by others, the remainder wait and see."[39] Despite the vigor of the discord internal to science, it should be remembered that what is being described here does not include the, at times, much more rancorous confrontations to First American science by the political activists representing the cultural equities of the American Indian community.

## Current First American Theories

Although there is certainly an intensity to the debate surrounding the search for the First Americans, the general commotion centers on relatively few theories. In other words, there are multiple avenues of attack upon each theory rather than a moderate number of challenges to each of a range of theories. Regardless of the specific science that may be employed, the noteworthy theories have separated into three domains.

First is the specific and the once long-reigning theory called *Clovis-first*, which will be discussed in detail in chapter 3. There is no doubt of the existence of the Clovis culture in North America. Rather, the argument concerns whether or not Clovis was the first occupation of the New World, with the alternative theories for an earlier arrival diverging by only a few thousands of years. The second set of contending theories addresses the geographical path by which the First Americans—whether Clovis or not—came to the New World; did they come via a Bering land bridge at the end of the last ice age, or did they come via a coastal route? Third, although most evidence indicates that the First Americans came from Asia, reputable scientists continue to develop theories that include a European origin for the First Americans—to be distinguished from the ancestors of modern American Indians who might well have migrated from Asia, but might also not be descended from the First Americans.

Occupying primacy of place for contentiousness as well as for importance to the deliberations has been the Clovis-first theory. From 1930 to 1999 (and for a few still today), the reigning First American theory was that thirteen thousand years ago the New World was occupied in a single migratory wave from Asia across the Bering land bridge that dashed all the way to Tierra Del Fuego in fewer than one thousand years and, in the process, hunted to extinction all of the indigenous megafauna including saber-toothed tigers, mammoths, and giant sloths. As befits a long-standing theory that is in the process of being rejected, Clovis-first has been attacked from many directions. The most substantive path of invalidation has been to show that Clovis-first was wrong because the Clovis culture was not actually first. Artifacts from the Monte Verde site in Chile have now been shown to predate the Clovis culture. The second attack on Clovis-first has been to challenge the megafauna extinction theory. While not critical to the Clovis-first model, hunting the megafauna to extinction fits a sort of bravado that was inherent in the Clovis-first model that painted the first inhabitants of the New World as being so vigorous as not only to quickly conquer an entire hemisphere by creating a new form of Paleolithic technology, but also to dominate the animal world that it found there. The third avenue of attack, from within as well as external to science, has been to challenge that the First Americans came via the Bering land bridge. Alternative maritime routes have been suggested for both the east and west coasts. It is possible that the Clovis culture could have come via a coastal route, but that was not a part of the Clovis-first paradigm. Finally, a few reputable First American scientists believe that the New World was settled via a maritime route by descendants of an ancient European Solutrean culture.

## Beringia

During the Cold War, many Americans became acutely aware that the Soviet Union was only a few miles across the waters of the Bering Strait from Alaska. Such a small separation achieved a comedic iconic status when in a September 11, 2008, interview with ABC's Charlie Gibson, then vice presidential candidate Sarah Palin stated that "you can actually see Russia from land here in Alaska."[40] The proximity, but physical separation, of the New and Old Worlds is a modern stabilized fact, what might be termed common knowledge. Yet geologists believe that as recently as ten thousand years ago toward the end of the last ice age, a person could have traveled by land eastward from Europe, traversed Asia, and crossed into the Western Hemisphere without being conscious of having changed continents.

The United States National Oceanographic and Atmospheric Administration (NOAA) maintains a website that provides voluminous information for nontechnical readers describing the glacial and interglacial cycles for the past 2.6 million years, noting that the "most recent glacial period occurred between about 120,000 and 11,500 years ago."[41] There was an abrupt cooling fifteen thousand years BP (called Last Glacial Maximum), which opened the Beringia land bridge for potential migration between Siberia and Alaska, followed by an equally pronounced warming (called Younger Dryas) ending 11,500 years ago after which the Beringia land bridge disappeared. During the maximum extent of the glacial ice, sufficient water had been absorbed into the glaciers that the Pacific Ocean was as much as 400 hundred feet lower that it is today, exposing between Siberia and Alaska a land bridge that geologists have termed Beringia.

Until recent decades, First American scientists discovered material evidence supporting human occupation of the New World only within the last fifteen thousand years; as a result, they tended to focus on the migratory availability of the Beringian land bridge only during the Last Glacial Maximum. However, with the alternating expansion and contraction of the glaciers in the Northern

Hemisphere, there would have been several exposures of the Bering land bridge over the past 100,000 years. The archaeologist Stuart Fiedel compared estimates by four scientists as to the availability of the land bridge for migration between Asia and North America and concluded that there were no fewer than six land bridge availabilities over the past 100,000 years.[42] Although such an analysis might not refute either the Clovis-first model or an initial migration via Beringia, it does provide a basis for the creation of alternative models on which a successful search for disconfirming material evidence might be sought.

Regardless of when a Bering land bridge migration might have occurred, as Adovasio acknowledged, "During Paleolithic times, migrants would have been insensitive to the act of crossing from one continent to another whether via Beringia or along its southern coast."[43] As straightforward (and straitforward) as the glacial history may be, there is significant contention as to whether the First Americans did indeed arrive in the New World via the Beringian land bridge. That some early migration could well have occurred by this foot path is generally not contested. What is debated is whether the First Americans arrived in this manner. The challenges come from those: (1) in the American Indian community who believe their ancestors were here from the beginning of time; (2) who believe that the First Americans came from Europe; and (3) who support a coastal (maritime) migration from Asia.

Criticism of the Beringia-first model from some parts of the American Indian community is not gentle. When commenting on the Kennewick Man debate, Deloria described the idea of the First Americans as having arrived in the New World after a migration across the Bering Strait as "a myth with little to recommend it."[44] In *Red Earth, White Lies: Native Americans and the Myth of Scientific Fact*, Deloria also commented that the Bering Strait theory "has been around so long that people no longer feel they have to explain or defend it—they can merely refer to it."[45] For Deloria and other American Indian activists, the scientific thesis of a primal migration via Beringia does not reflect tribal memories and traditions. Some of these traditions do mention boats and maritime migrations. "The Hopi and Colvilles, for example, and others speak of the experience of a creation, such as the Yakimas and other Pacific Northwest tribes. Some tribes even talk about migrations from other planets."[46] It is not difficult to understand why scientists have not accommodated within their theories an interplanetary origin for the First Americans. However, Deloria is quite correct that science has generally accepted migration across Beringia as a virtual material artifact as opposed to a theory, which is itself a social construction built on truly material evidence. As Jace Weaver, director of the Institute of Native American

Studies and professor of religion and law at the University of Georgia, pointed out, "The Bering Strait theory is simply that—a theory."[47] Although theories such as polygenesis and interplanetary travel may appear outlandish, Weaver contended that the freedom to consider such theories "is vitally important in freeing the imagination and opening up the seams of what has been largely a closed discussion. . . . The truth is nobody knows the origins of the Native tribes on this continent."[48] Nevertheless, not only within science but also in popular American culture, Beringia as a source of ancient migration to the New World has become a fact, a culturally stabilized bit of American knowledge.

There is criticism of the Beringia land bridge model from within the First American scientific community as well. An excellent example is the 2006 PhD dissertation by Thomas George Arnold at Simon Frazer University in Canada. Based on an examination of six hundred published radiocarbon dates, Frazier devoted 263 pages to answering two questions: (1) could eastern Beringia (what is now Alaska and western Canada) have supported a migratory population immediately prior to the appearance of the Clovis culture; and (2) did a biogeographic corridor through the ice sheets exist prior to 11,500 BP, thirteen thousand chronological years, that could have supported a north-to-south migration? Arnold concluded that Beringia could indeed have hosted a human population prior to the appearance of Clovis and there could, therefore, presumably have been an antecedent culture. However, he also concluded that "the ice-free corridor could not have been used as a north-to-south human migration route . . . [and] other alternatives must now be considered to account for the arrival of Paleoindian cultures in southern North America."[49] Arnold's work echoes that of Carol Mandryk, Dennis Stanford, and Bruce Bradley in challenging the dominant only-Beringia theory.[50] Although their alternative theories certainly cannot be considered a definitive refutation of the thesis that Beringia was the migratory path of the First Americans, they do demonstrate that its status as a *fact*, as a bit of stabilized knowledge, appeared to be unwarranted.

The artifactual dating associated with the Clovis model serendipitously matched the timing of a possible entry into North America via Beringia when some First American scientists believe there would have been land routes passable through the receding glaciers. Consequently, leading scientists have frequently taken a dismissive attitude toward potential evidence that entry may have actually been sea-borne. With the recent challenges to the Clovis-first model, renewed attention is being paid to the coastal model. For example, human remains at Arlington Springs on Santa Rosa Island off the coast of California appear to predate Clovis.[51] Since the obvious maritime implications of an island archaeological site cannot be challenged, in such an instance protection of the Clovis-first paradigm was typically manifested via challenges to the dating techniques. In recent years, the hypothesis of coastal migrations has seen additional supportive evidence. In 2018, Florida state officials formally announced the discovery of human remains in a seven-thousand-year-old submerged burial site off its coast. In the same years, archaeologists announced the discovery of thirteen-thousand-year-old human footprints on an island off the coast of British Columbia. As difficult as it is to locate ancient human remains on land, it is extremely difficult to find those that are submerged; archaeologists must first have some notion as to where to investigate, must have the appropriate maritime skills and equipment, and must also serendipitously locate remains that have not been irretrievably damaged by the wet environment. But, as some First American scientists seem to have overlooked, absence of evidence is not necessarily evidence of absence.

## Migration Modeling

The successful practice of First American science relies increasingly upon mathematically based skills located outside of the fields of archaeology and anthropology. For example, independent of any particular First American theory are techniques involving computer simulations and stochastic modeling that can be applied to human demographics such as migration dynamics, reproduction success, and technological innovation. As pointed out in 2000 by the archaeologists David G. Anderson, at the University of Tennessee, and J. Christopher Gillam, then at the University of South Carolina, there had been relatively few continental-scale studies undertaken to determine how landscape features such as rivers, mountains, passes, and lake shores might facilitate or, alternatively, impede the progress of colonizing peoples.[52] Migratory-path modeling, especially when accompanied by biogeographical analysis such as that referenced earlier by Thomas George Arnold, can be a powerful tool for bounding feasible First American theories.

As Meltzer has acknowledged, "there are subtle assumptions in this dispute that might constrain the manner in which the problem is approached or the evidence is evaluated."[53] Meltzer then proceeded to list some of these important, and largely unsubstantiated, suppositions including: Hrdlička's 1926 belief that the First Americans were "dribbling over from northeastern Asia, extending probably over a long stretch of time"; Paul Martin's 1973 view that the earliest migration resulted from a "one-time event involving 100 individuals"; C. Vance Haynes's 1987 contention that evidence of the arrival of the First Americans should "logically lead" to the Clovis culture; and R. E. Morlan's 1987 assumption that the first Americans were a product of "arrow-like purposeful migrations."[54] Meltzer's retrospective look at these specious assumptions embedded in the works of prominent twentieth-century First American scientists can only embolden the critics, both internal and external to science, of such theories as Clovis-first and migration across Beringia. Especially egregious is a conclusion that the migration was *purposeful*. Migrations by Moses or the Pilgrims could conceivably be described as purposeful since there is a written record purporting to reflect their motivational intent. However, based on scant archaeological evidence, it is difficult to comprehend how the psychological state of a prehistoric population could accurately be judged by a First American scientist separated from them by thirteen millennia.

The use of statistical modeling can facilitate identification of assumptions that might, in hindsight, be as questionable as some of those cited above. However, it can also be used to determine the feasibility of models that might otherwise appear counterintuitive. The goal of Anderson and Gillam was to identify the paths, dispersal rates, and geographical logic behind the initial population migration in an attempt to provide boundaries within which First American theories could reasonably be constructed.[55] To do this, they postulated two different migration models that were named for the visual patterns created when their movements were displayed across a map of the New World. The first model, called "string-of-pearls," assumed that once a population grew to a sufficient size, a subgroup splintered and moved into an adjacent territory. The second, or "leap-frog," model conjectured that the splintering group moved a significant distance from the founding population. In both scenarios, this process provided a continuing geographical expansion of the population.[56] According to the string-of-pearls scenario, it would require over five thousand years for colonizers arriving through Beringia to reach the southern tip of South America; this period would be reduced to approximately four thousand years if the population took a maritime route down the west coast of the hemisphere. The leap-frog model, however, dispersed the population much more quickly over the terrain, and required fewer than one thousand years to range from Alaska to the southern tip of South America—which makes feasible the rapid expansion required by the Clovis-first model. Moreover, according to Anderson and Gillam, "the distribution of [Clovis] fluted points in the lower 48 states in dense but widely separated clusters, many occurring in resource rich locations, coupled with a more widespread low density distribution, in fact, suggests early populations did use a leap-frogging movement strategy."[57]

Clearly, there is value in the application of statistical models that can help establish credible boundaries for the creation of First American theories. For example, there is a certain intuitive disbelief associated with a Clovis-first model that required Paleolithic colonizers in a mere one thousand years to populate the entire Western Hemisphere on foot, using stone tools, crossing formidable geographic barriers, while continuously adapting to new landscapes with significant climatic changes. However, using reasonable assumptions, Anderson and Gillam demonstrated that the rapid expansion required by the Clovis-first model was not impossible.

## Solutrean

Not all scientists are convinced that the First Americans, to be distinguished from modern American Indians, were from Asia. A few reputable First American scientists, such as Bruce Bradley and Dennis Stanford, contended that the first route of entry was by Europeans via the east coast and that the First Americans were the descendants of the twenty-thousand-year-old Franco-Cantabrian Solutrean culture, which originated in the Paleolithic Iberian peninsula and came across the Atlantic Ocean to first populate America. Although Bradley and Stanford did not dismiss Beringia as providing a possible path to the New World for later migrations, their argument challenged Beringia's importance since it necessitated acceptance of a European origination as well as a maritime route across the diminished expanse of the North Atlantic during the last ice age.[58] While this is a distinctly minority view, the Monte Verde dislodgement of the Clovis-first theory has opened avenues for reconsideration of contradictory evidence. For example, stone tools found at the Cactus Hill, Virginia, and Page-Ladson, Florida, sites do not resemble the Clovis tool kit, appear to have no antecedents in Siberia and Beringia, but have similarities to those of the ancient Solutrean culture in Europe.[59] There are obvious political considerations outside the scientific community that compound the difficulty in this model gaining acceptance within the community itself. On the one hand, there are elements in the public that could be expected to support a European model for racist reasons. On the other, there are also those who may reflexively discount the Solutrean model without directly addressing its evidentiary merits in order to preemptively preclude any appearance of racism. From time to time—more often in the public press, at least blatantly, than in the scientific community—racism is a factor in the discussions, as was evident in the press reports on the remains of the Kennewick Man.

The Upper Paleolithic Solutrean culture is a particular collection of stone tools documented to have been in southwestern France and Iberia in the period ranging from 25,000 to 19,700 years ago—hence the mantra of "Iberia not Siberia" that is offered by Solutrean proponents in countering their colleagues who espouse an Asian origin. Some archaeologists believe that not only can the Cactus Hill and Page-Ladson artifacts be tied to an antecedent Solutrean origin, but also many of the distinctive characteristics of Clovis material artifacts can be linked to a Solutrean origin. Moreover, the oldest and most productive Clovis sites are in the southeastern United States, and not, as might be presumed, in the western United States and Canada nearer the Bering land bridge, which many scholars contend was the path of the original migration that produced the Clovis culture. In support of the Bradley and Stanford hypothesis, during the last ice age, the uncovering of the continental shelves meant that Europe was separated from North America by only 1,400 miles broken by occasional land and glacial islands.[60]

The thesis of a European Solutrean origin, however, is not without its difficulties. The most obvious is that, with the exception of rare European-looking discoveries such as that of the Kennewick Man, there is little fossil or genetic evidence that archaic Europeans, and specifically the Solutreans, roamed the New World. Moreover, even if it could be conclusively proven that the Solutreans were indeed an antecedent culture to Clovis, that would only confirm them as even earlier Americans, but not necessarily the First Americans—precisely the problem that plagued the Clovis-first theory. Finally, as James Chatters and others have pointed out, the Clovis culture arose approximately 13,500 years ago; yet the Solutrean culture was supplanted in Europe by the Magdalenian culture about 19,700 years ago. So, where did the Solutrean culture disappear for six thousand years until it purportedly reestablished itself in the New World as the Clovis culture?[61] If it arrived in the New World twenty thousand years ago, where is the evidence? These questions have yet to be answered.

## Conclusion

The Clovis-first model was the initial theoretical product of modern First American science, which, based on a marriage of material evidence and modern scientific methods, offered a comprehensive explanation for the First American origin (Siberia), physical entry path (Beringia), cultural tool kit (Clovis fluted points), area settled (entire Western Hemisphere), and environmental adaptation (hunted megafauna to extinction). It bested all challenges for seven decades and was accorded the respect of presumptive veracity that a reigning paradigm enjoys. Clovis-first provided the foundation that stabilized scientific as well as popular knowledge associated with the search for the First Americans. The problem, however, is that many First American scientists now believe the Clovis-first model was wrong. The excavations at Monte Verde, Chile, have shown that there was indeed at least one New World culture that predated Clovis, with an increasing expectation that other sites—such as Meadowcroft Rockshelter, Page-Ladson, Cactus Hill, Buttermilk Creek, and Chiquihuite Cave—offer additional pre-Clovis challenges. Unfortunately for those who support a pre-Clovis arrival, there are no associated comprehensive theories. First American scientists now know that other culture(s) existed before Clovis, but there is scant material evidence as to where they came from or how they got there upon which to build a broad explanatory theory.

# CHAPTER 3

# Clovis-First

IN 1906, THE SANTA FE RAILROAD operated a small station in eastern New Mexico called Riley's Switch. As a part of the search for an official name for the town that was developing around it, the daughter of a railroad official was permitted to suggest a name. Captivated by her study of a fifth-century Frankish king who had become Christian, the daughter suggested Clovis. In May 1909 the town of Clovis was officially incorporated.[1] As a result, one of the first indigenous cultures of the New World has been irretrievably encumbered with a name prominently associated with an alien European culture.

The theory against which all other First American science was measured for seventy years was the "Clovis-first" paradigm. In 1929 at Blackwater Draw near Clovis, New Mexico, the material artifacts that became the Clovis cultural "tool kit" were unearthed and the culture was subsequently successfully dated to thirteen thousand years ago—the then-oldest documented evidence of human presence in the Western Hemisphere. The stone artifacts have a shape that is characteristic of a culture that purportedly existed only in the New World, and only for a limited period of time.

Blackwater Draw, New Mexico—the physical location that produced the original Clovis artifacts—is actually located some fourteen miles southwest of Clovis. The closest town, albeit a small one, is Portales, New Mexico, which is six miles still farther south. In 1929, local resident Ridgely Wightman discovered and sent to the Smithsonian Institution a piece of mammoth bone and a stone point that he found in Blackwater Draw. Consequently, Wightman, a non-scientist, can be credited with the discovery of what became known as the Clovis culture.[2] According to the October 20, 1932, *Portales Valley News*, "many prehistoric bones are being uncovered at the highway gravel pit northwest of Fort Palace and there is an interesting display of these bones in the show window of Ed J. Neer's store. Centuries ago monsters roamed the prairies where Portales now stands, and one can visualize the strange animals after looking at the bones now being uncovered."[3]

From its beginning, the artifacts discovered at the Blackwater Draw site have demonstrated the unique intersection that the search for the First Americans has created among science, popular public interest, American Indian myths, and governmental politics. The site was designated by the US government as a National Historic Landmark, and a cottage industry has developed around it. In early 1933, E. B. Howard, of the University of Pennsylvania, and John Cotter, of the Philadelphia Academy of Natural Sciences, reported the first successful results by professionally trained investigators.[4] That same year a Clovis expedition party with members of the prestigious International Geological Congress included R. Prall, Victor Van Straelen, Peter Anderson, Lady Smith Woodward, Sir Arthur Smith Woodward, Chester Stock, John C. Merriam, Jake McGee, R. W. Wilson, Ridgely Whiteman, Edgar B. Howard, and either F. D. Bode or H. D. Curry.[5] A participant in that excavation commented some years later that "We excavated industriously, in our quest for mammoth bones and artifacts. Not quite as modern archaeological practice requires, but not without method."[6] Excavation continues to this day under the auspices of Eastern New Mexico University.

After scientists knew what to look for, variations of this Clovis tool kit (principally associated with a lanceolate fluted projectile point, scrapers, and ivory, antler, and bone implements) were found throughout North America. Despite the absence of convincing evidence of these tools having been found in South America—based largely on an erroneous belief that Clovis was older than any found in South America—the Clovis-first model contended that the Clovis people crossed the Beringian land bridge from Asia during the latter part of the last ice age, very quickly spread through North America, and rapidly populated the entire southern continent as well. The stature of the Clovis theory became such that any archaeological evidence hinting at an earlier peopling was termed *pre*-Clovis—not worthy of an independent designation—and its scientific value usually discounted. Associated with the Clovis model were assertions about the number of subsequent migrations from Alaska, the precise route of entry (between or around glaciers), and whether the Clovis people were responsible for the extinction in the Western Hemisphere of megafauna such as the mammoth and the mastodon. With the near-consensus of the past two decades that archaeological evidence from Monte Verde, Chile, disproves the Clovis-first model, there is now no clear theory explaining who the first Americans were and by what route they entered the New World.

The Clovis-first paradigm had been able to offer both archaeological sites (observations) as well as an associated explanatory model (theory) in its support. While there are several sites—some are more contested than others—that are now believed to predate the Clovis sites, there is currently no associated explanatory theory to support their existence. The Thomas Kuhn thesis that one paradigm cannot be overturned, even in the face of contradictory evidence, until there is a suitable theoretical replacement would appear not to be sustained in this instance. The single most important pre-Clovis site is that of Monte Verde II. Its antiquity (fourteen thousand years ago), location (Chile), and rigorous archaeological documentation have shown the Clovis-first model to be untenable. Other important pre-Clovis sites include Meadowcroft Rockshelter (Pennsylvania), Page-Ladson (Florida), Cactus Hill (Virginia), Buttermilk Creek (Texas), Chliquihuite Cave (Mexico), and Paisley Cave (Oregon). In 2017 the San Diego Natural History Museum announced that it had discovered artifacts documenting a human presence in California dating to 130,000 years ago. This has yet to be confirmed by the broader First American community in what will be a contentious endeavor.

In 1990, a *Science* article entitled "Clovis Counterrevolution" described the, at times, vicious debate between the Clovis and pre-Clovis First American science communities.[7] Indeed, the major academic figures could generally be placed into one of two camps. The most prominent people in the Clovis-first intellectual encampment—characterized in *Science* as the "Clovis police"—were: Thomas Lynch of Cornell University, who is generally credited with primacy of place in the development of the Clovis-first theory; Dina Dincauze of the University of Massachusetts; Paul Martin of the University of Arizona; and C. Vance Haynes, also at the University of Arizona.[8] Haynes is generally credited with having been the leading proponent providing the continuing force behind the Clovis-first theory. Some, such as Karl Butzer, an archaeologist at the University of Texas, believed that the adamant skepticism of the Clovis-first supporters, against countervailing theories, may well have become a cult that dismissed "any kind of evidence, regardless of what it is."[9] Ten years ago, the phrase "Clovis police" was so commonly applied to the protectors of the Clovis-first theory that numerous entries were listed on an internet search, even after modifying it to "Clovis police archaeology" to eliminate references to the Clovis city police department. Given the shift away from the Clovis model, today the archaeological politics on the internet have been considerably muted.

Those supporting a pre-Clovis arrival in the New World, and who frequently were associated with viable specific pre-Clovis archaeological models, most prominently have included Tom D. Dillehay of Vanderbilt University (Monte Verde site in Chile); James Adovasio of Mercyhurst College (Meadowcroft Rockshelter in Pennsylvania); and Dennis Stanford of the Smithsonian

National Museum of Natural History (precursor Solutrean culture from Europe). Also included in the early pre-Clovis group were those who received less academic and popular press coverage in the United States because their research is centered outside the United States, including Alan Bryan and Ruth Gruhn of the University of Alberta in Edmonton (Taima-Taima in Venezuela) and Jacques Cinq-Mars of the Canadian Museum of Civilization (Blue Fish Caves in Canada). There have been, however, a few leading First American scientists who may have had tendencies toward one theoretical commitment or the other, but whose public expressions could be better described as academic statesmanship rather than consistent support for any particular position. One of the prominent few who fell into this mediator role has been David J. Meltzer at Southern Methodist University.

The focus of the First American debate has typically centered on whether, when, and how a migration occurred at the end of the last ice age from Siberia to the New World by means of a land bridge called Beringia. However, as the genetic evidence indicates, a Beringian migration may well have been both ways. Reflecting the popular American interest in the search for the First Americans, an August 2, 1996, *New York Times* article titled "'American' Arrowhead Found in Siberia" proclaimed that the archaeologists Sergei B. Slobodin and Maureen L. King had discovered a Clovis point in Siberia that radiocarbon dating indicated was only 8,300 years old. That would place it in Siberia some two thousand years after the demise of the Clovis culture in the New World.[10] Since that report by Slobodin and King appeared, however, the existence of Clovis-type cultural artifacts in Siberia has generally been discounted.

One of the more romantic notions associated with the Clovis theory is that not only were these occupiers of the New World the first to arrive, but as they sped from Alaska all the way to Tierra del Fuego, they were such consummate masters of their domain that they hunted to extinction all megafauna, generally considered animals larger than humans. What has not been answered, however, is that if Clovis sped so quickly throughout South America, why later did the much more culturally and technologically advanced Aztec and Maya civilizations not also expand to larger areas? Although Carl Sauer is credited as having proposed seventy-five years ago that early man had killed off the North American big game animals through the use of fire drives, Paul Martin is generally recognized with proposing in 1967 the theory that the Clovis hunters caused the extinction of the megafauna in his 1967 *Nature* article titled "Pleistocene Overkill."[11]

While there is still support in some quarters for Martin's thesis, today it has generally been discounted. Martin documented that the rise of the Clovis culture and the demise of the megafauna occurred during approximately the same period, but never successfully identified a causal relationship. As Adovasio summarized the critique, for the Clovis-first theory to be valid, it would be necessary for a small band of as few as twenty-five male super-predators to have conducted a blitzkrieg consisting of: penetrating through the glaciers of North America; inventing the Clovis stone tools; covering six thousand miles in possibly as few as five hundred years to reach Tierra del Fuego; not only surviving but reproducing at the astonishing rate of 3.4 percent annually; adapting almost instantaneously to a broad range of new ecological challenges; and driving to extinction across one-fourth of the world's land mass all animals that weighed more than one hundred pounds.[12] That such seemingly implausible theories can be put forward by reputable modern scientists lends credence to the arguments of those such as Deloria who contend that American Indian myths are as substantively explanatory of the physical world as are First American scientific theories.

In addition to the conjecture that the Clovis culture was responsible for the disappearance of the mammoths and saber-toothed tigers, there is even more that has been associated with the Clovis theory that caters to the popular imagination. There is not only the question of Clovis cultural beginnings, but also uncertainty as to its demise. Just as it arose swiftly, it also disappeared swiftly. In 2006, Lawrence Berkeley National Laboratory geologist Richard Firestone theorized that a comet struck the Earth 12,900 years ago and led to the death not only of the Clovis culture, but also to the extinction of the megafauna. While it may be hard to resist the romantic notion that the demise of the First Americans, mammoths, and saber-toothed tigers could only be produced through extraterrestrial intervention, Meltzer and Vance Holliday of the University of Arizona provided a detailed rebuttal of the comet theory.[13]

A principal reason behind the intense skepticism surrounding pre-Clovis sites is that there have been many such sites thought to have been discovered that were later properly dismissed for lack of sufficient evidence, such as Tule Springs, Nevada; Holly Oak, Delaware; and Calico Hills, California, where Louis Leakey famously erred.[14] Although such a dismissive approach may have some intuitive comfort, it is difficult to understand how an academically trained scientist can contend that scientific observations in one case are either stronger or weaker because of unrelated observations in another instance made at a different time and a different place. This bias in First American science, according to *Science*, "dates back to the tenure of Aleš Hrdlička and William Henry Holmes at the Smithsonian Institution in the 1920s," and continues today from a self-interest in preserving the status quo.[15]

As with any major scientific endeavor, the search for the First Americans has produced its share not only of worthwhile endeavors, but also a certain number of mysterious undertakings. One example reminiscent of an Indiana Jones movie is that of an Argentine who excavated for nearly three years at Blackwater Draw and left no notes. In 1967, Alberto Isequilla, believed to have then been a graduate student at either the Sorbonne or at the University of Paris, arrived at Blackwater Draw and began a major excavation project.[16] So much earth was removed that the area became known as Isequilla's Pit. By 1969, after digging his enormous hole, Isequilla "just left."[17] Isequilla was later reportedly living in France, possibly as an art dealer and as CEO of d'Arts Finans Trust involved in the expanding art market for Russian oligarchs.[18]

After the 1969–70 excavation season, a plastic cover was placed over the pit. In 2009, the overburden that had accumulated over the forty-year period was removed. The Eastern New Mexico University (ENMU) Archaeological Field School subsequently used the site as a training area for excavations. According to the anthropologist David Kilby, "Little is known of the extent of excavations or the nature of what was encountered, and Isequilla's 1967–69 fieldwork remains a poorly known episode in the history of research at the Blackwater Draw site."[19] Isequilla reportedly did not finish his dissertation, and his field notes have disappeared. "We have this depression out here and a hole that he filled in with sand. No one knows, for sure, exactly what he excavated, how deep they got or what the results were."[20]

Amid the challenges mounted against the Clovis-first model, the Clovis-associated megafauna extinction, and the Clovis extraterrestrial demise theory, it must be remembered that there is no meaningful dispute among First American scientists that Clovis was indeed a very early New World culture, created a unique set of material tools, and populated much of North America. Clovis, however, is also important as an example of a form of scientific practice that creates a theory that rests upon a particular set of material evidence, and then denies the legitimacy of other evidence because it does not conform to the rigidified theory. The absence of contradictory evidence when a paradigm is created does not necessarily provide any insight as to whether such evidence may ever exist. In the case of the Clovis culture, the Clovis-first model was considered so "validated" in the minds of many First American scientists that it was granted a concreteness that surpassed the physicality of those artifacts that directly challenged its validity. As Meltzer concluded, "The assumptions that the earliest migration involved a single, homogenous population and that pre-12,000 BP occupations must be ancestral to Clovis, are not well-founded. . . . All this, in turn, has implications for why we may not know when the first people came to North America."[21]

Discussion of the settlement of the New World has been rancorous in the First American science community. Moreover, it has occurred simultaneously with the relatively peaceful approach to establishing the human migration into Australia.[22] As Arthur J. Jelinek, an archaeologist at the University of Arizona, has pointed out, "This situation [the Clovis versus pre-Clovis controversy] stands in strong contrast to that in Australia, where, in less than 30 years, many fewer archaeologists than have been active in New World Paleoindian research have produced undisputed evidence of a human presence beyond 30,000 BP in widely separated regions, and strong indications of human activity even 10,000–20,000 years earlier.... There are now well over 30 sites known in Australia with dates greater than 17,000 years, and over half a dozen with convincing dates in excess of 30,000 years."[23] Yet, there is a natural inclination to compare the First Australian and First American searches since Australia and the Western Hemisphere are the only continental landmasses for which archaeological investigations have been conducted on recent colonization by anatomically modern human beings.[24] If the First Americans arrived prior to fifteen thousand years ago, they left no archaeological record as did their Australian counterparts who arrived forty thousand years ago. If the First Americans simply "disappeared after leaving a few ephemeral traces of their presence to confound the archaeological community ... then they were merely a historical curiosity and of little import or interest for the cultural development of the New World."[25]

# CHAPTER 4

# Pre-Clovis: Monte Verde and Meadowcroft

T HERE ARE SOME ELEMENTS of the Clovis model that appear beyond challenge. It is impossible to dispute the existence of the Clovis cultural tool kit, and it must also be acknowledged that the Clovis culture is the first that has been found with a widespread dispersal across the North American continent. What can be, and indeed has been, alleged is that the Clovis culture never arrived in South America, that the Clovis civilization did not lead to the extinction of the megafauna in North America, that the American Indians are not descendants of the Clovis people, and that the people who produced the Clovis culture were not the First Americans. It is this final contention, that there were pre-Clovis cultures, that generated considerable acrimony among First American scientists. Two prominent pre-Clovis sites that initially challenged the Clovis-first model were Monte Verde in Chile and Meadowcroft Rockshelter in Pennsylvania.

## Monte Verde, Chile

One of the enduring debates in the philosophy of science is the (im)possibility of demonstrating a conclusive proof for any particular scientific theory. Indeed, there is a significant body of thought based upon the work of Karl Popper that holds that for any theory to be considered scientific, it must potentially be able to be falsified; that is, there must be some conceivable test that, if successfully conducted, would disprove the theory.[1] Although repeated instances of successful excavations of Clovis sites would reinforce the importance of the Clovis culture, it only takes one instance of a successful excavation of a pre-Clovis site to disprove the Clovis-first model. In this case, the necessary single instance of disproof was the excavation at Monte Verde, Chile, during the last three decades of the twentieth century.

The Clovis-first theory enjoyed the appearance of being confirmed by a broad and deep material record. However, in retrospect, the theory was composed of an amalgamation of largely independent theories that can best be termed fellow travelers—the ultimate success of any particular subordinate theory was neither dependent on the success of the Clovis-first theory itself nor dependent on any other subordinate theory. The relationship can, perhaps, be described as a political accommodation. Some First American scientists were aware that the parts from which Clovis-first was assembled had a tenuous relationship. Meltzer noted that a successful challenge to the Clovis-first theory need only successfully demonstrate a pre-12,000 BP material record; it need not specifically disprove such associated theories as a Beringia crossing or a human-driven extinction of the megafauna. According to Meltzer, "the question of whether people were present at a given time must be kept separate from more complex questions about how they lived."[2] Nevertheless, the burden that Monte Verde now carries is that, while it has displaced Clovis as the earliest documented human occupation in the New World, there is no commonly accepted associated theory describing where those people came from, what path they took, and the extent to which their descendants survived in the New World. Within this theoretical vacuum, there is little reason to believe that Monte Verde will be immune from having attached to it an assortment of fellow traveler theories whose success will be tied to the success of Monte Verde.

The Monte Verde excavation site is a few hundred miles south of Santiago, Chile, near Puerto Montt and next to Chinchihaupi Creek. In what has become a typical scenario of a significant archaeological discovery, in 1975 some old bones were serendipitously discovered by a non-archaeologist. Archaeological excavations began in 1976 under the auspices of Tom D. Dillehay and Mario Pino of the Southern University of Chile in Valdivia. Dillehay is now at Vanderbilt University. Because the site had been covered by a water-saturated peat bog, the deterioration of normally perishable materials such as wood and textiles was greatly reduced because the water prevented the decay-inducing effects of oxygen.[3] In the search for the first Americans, the implications of the Monte Verde excavation are profound. Not only is the human occupation at least one thousand years older than the Clovis culture, it is ten thousand miles from what was the potential colonization route in Beringia.

There are two levels of excavation at the site. The upper, or younger, level is designated MV-II with an estimated age of 14,500 BP. As discussed below, it is at this level that the material evidence refutes the Clovis-first model. Nevertheless, it does not necessarily invalidate all of the associated subtheories associated with Clovis-first. For example, First American scientists are not able to precisely determine the dates during which humans might potentially have crossed Beringia and penetrated the ice fields of North America, nor able to confirm or refute the megafauna extinction theory. Although the dates for MV-II are sufficiently earlier than Clovis to refute Clovis-first, the chronological difference is not so large as to require a coastal penetration scenario.

While acceptance of MV-II alone is sufficient to disprove Clovis-first and to require a rewriting of the First American scenario, the possible implications of Monte Verde are even greater. Beneath the level of MV-II, Dillehay has found material evidence of potential human-associated activity at a level, termed MV-I, initially dated at 33,000 BP and later revised to 18,500 BP. There has been no finally accepted report for MV-I. As leading First American scientists have termed it, the status of MV-I "remains unresolved."[4] If, however, the MV-I findings are validated, it will require a completely new script for human colonization of the Western Hemisphere. It will be a direct challenge to Beringia, megafauna overkill, and Siberia as subtheories for the First Americans. "The chances seem good that these materials indicate a significant early human occupation in the region."[5] Pending publication and acceptance with regard to the MV-I excavation, the focus of First American scientists understandably remained concentrated on the implications of MV-II. The principal human-occupied area excavated at MV-II has now been destroyed by a meandering stream, logging activities, and road construction.[6]

Because of the historically intense—frenzied might be more appropriate—scrutiny of any challenge to the Clovis-first model, Dillehay did not publish his final report associated with MV-II until twenty years after he began the excavation. In First American science, the boundary between what constituted a material fact and its associated explanatory theory seemed to have disappeared, with the Clovis-first model giving every indication of having been honored with the status of fact, rather than explanatory theory, by many in the First American scientific community. The material artifacts associated with MV-II in no way challenged the validity of the material artifacts related to the Clovis-first theory. However, the material artifacts associated with MV-II absolutely challenged the Clovis-first theory itself.

The difficulties in an archaeological excavation lie in two areas. The first is a determination of a human presence, frequently in the absence of any human biological remains. Consequently, the evidence is typically composed of associated materials that reflect human activity. Documenting this activity is more difficult than might be imagined, since natural and animal-related activities are capable of fracturing stone and marking wood or bone in a nearly perfect replication of a human effort. Once a human-related activity has been determined, the second challenge for First American scientists is to chronologically situate the human presence. The overwhelming majority of the Clovis sites had no direct evidence of human presence in the form of skeletal remains; consequently, the significance of the sites was based on evidence of human-related activity in the form of artifacts. However, at Monte Verde, Dillehay was in the envious position of having direct evidence of a human presence in what was judged to be a clear human footprint that was preserved in the peat bog. Once a chronological determination was made—and supported by other evidence of human activity such as tent pegs, tools, and bits of woven basketry—Dillehay offered to First American scientists a sufficiency of evidence at MV-II to overturn the Clovis-first model.

There was not, however, an immediate and broad acknowledgment of the validity of Dillehay's findings. First American science is a community endeavor, and MV-II not only challenged the reigning paradigm of Clovis-first, but also confronted the authority of the elite members of the First American scientific community who had for decades actively supported a Clovis-first theory. After two decades of publishing shorter topic-specific articles, in 1997 Dillehay produced the long-awaited second volume of his research titled *Monte Verde, A Late Pleistocene Settlement in Chile: The Archaeological Context and Interpretation.*[7] Also in 1997, what *Archaeology* termed a "blue ribbon commission" of First American scientists conducted a site visit at Monte Verde.[8] In October 1997 this group, which included strong proponents as well as long-standing skeptics of the Clovis-first model, jointly authored a report in *American Antiquity.* They concluded by consensus that "the MV-II occupation at the site is both archaeological and 12,500 [BP or 14,000 chronological] years old, as T. Dillehay has argued."[9]

Despite the 1997 legitimizing of MV-II by the consensus of the elite of First American science, the caustic exchanges that are frequently characteristic of the academic community continued. Junius Bird of the American Museum of Natural History, who Jace Weaver called the "dean" of Paleoindian archaeology "joined in dismissing the discovery and attacking Dillehay."[10] In an October 1999 article, Stuart Fiedel attacked Dillehay's product as well as his professionalism.[11] In a December 1999 response by Dillehay and nineteen scholarly coauthors titled "On Monte Verde: Fiedel's Confusions and Misrepresentations," Fiedel was accused of a "misunderstanding of and inexperience with" research design and analysis as well as "factual and interpretative errors and misrepresentations."[12] With the unstabilized power relationships involved in the search for the First Americans, and with no consensus theory to replace the discounted Clovis-first model as to the colonization of the New World, pettiness appeared to rule. As Dillehay commented, "Instant-opinion-hurling has become something of a sport in the study of the first Americans—a sport that reveals our arbitrary understanding of . . . the peopling of the Americas."[13]

Balanced against the intramural squabbling among First American scientists, however, is a commitment in some quarters to what is advertised as the objectivity of science. As Meltzer has pointed out, it only requires one site to disprove Clovis-first. "Just as there is no compelling evidence to accept a pre-Clovis occupation, there is no compelling evidence to deny one either."[14] This proscription would not appear to capture how First American science has responded to challenges to Clovis-first. The difficulty, perhaps, is that while Clovis-first has been overturned, Monte Verde has offered no alternative explanatory theory. In the absence of a new explanatory paradigm, the attacks are frequently *ad hominem*— since there is no message, attack the messenger. While any number of other locations could have been examined here, the Meadowcroft Rockshelter provides a fertile example of the conflicts surrounding the challenges to Clovis-first.

## Meadowcroft Rockshelter, Pennsylvania

On July 13, 1974, James Adovasio, then a professor of anthropology at the University of Pittsburgh, received a radiocarbon-dating report from the Smithsonian Institution documenting ages of 12,975 ± 650 BC and 13,170 ± 165 BC from two different human-associated fire pits in Meadowcroft Rockshelter in what is now western Pennsylvania.[15] In a 1977 article early in the excavation process, Adovasio and his coauthors announced that "seven classes of artifactual remains" were discovered, including hundreds of items of "lithic, bone, wood, shell, basketry, cordage, and ceramic materials."[16] According to Adovasio, Meadowcroft Rockshelter "stirred up not only scholars but also many of today's Native American peoples by casting doubt on the legitimacy of their claim to be the descendants of the first Americans."[17]

The Monte Verde and Meadowcroft Rockshelter excavations share many common facets. First, excavations began in the 1970s at both locations as a result of serendipitous discoveries by amateurs. Second, the excavations were led by credentialed academics—Dillehay at Monte Verde and Adovasio at Meadowcroft Rockshelter—who were, and still are, recognized members of the archaeological community. Third, both excavations have been conducted in a decades-long environment of intense skepticism by First American scientists because of the challenges that each site presented to the Clovis-first theory. Finally, either alone was capable of profoundly affecting what had been well-confirmed theories with regard not only to the identity of the First Americans but also to the timing of their arrival.

Despite the congruence between these two projects, there has also been an area of substantive difference specifically related to the social context within which each excavation has been conducted. At Monte Verde, based on the material evidence that ultimately produced a successful challenge to the Clovis-first model, Dillehay's efforts conformed to the general mores of the First American science community. During the period in which he conducted his excavation and prepared his findings, he typically refrained from the personal pettiness that the search for the First Americans has seemed to generate. After twenty-five years of research and publication, in 1997 he fully cooperated with the inspection of Monte Verde by the elite of the First American science community. In addition, when Stewart Fiedel attacked his final report, as described above, Dillehay responded not alone, but with a brief article supported by nineteen coauthors, including many preeminent First American scholars. By comparison, Adovasio has made clear his disdain for the manner in which the search for the First Americans has been dominated by a select few who, according to Adovasio, have not been evenhanded in evaluating evidence that challenged the reigning Clovis-first paradigm.

To date, Meadowcroft has not been granted Monte Verde's status as having successfully challenged the Clovis-first model. A case might be made that this is at least partially due to Adovasio's difficult relationship within his own scientific community. However, a substantive reason is that, unlike Dillehay, he has yet to publish a detailed report on the results of his excavation. In 2003 he published a book titled *The First Americans: In Pursuit of Archaeology's Greatest Mystery* written for the popular press.[18] Although it was a summary of his work at Meadowcroft Rockshelter and an entertaining survey of the First American search, he not only failed to provide an academically rigorous presentation, but he also continued his pejorative insinuations about the First American scientific community. It would be an interesting historical case study to determine the extent, if any, to which the social dynamics between Adovasio and the preeminent First American scholars played a significant role in either delaying the demise of the Clovis-first paradigm or in Meadowcroft Rockshelter not being initially acknowledged as a legitimate pre-Clovis site.

Regardless of his relationship with First American scientists, Adovasio has been effective in garnering support outside of the scientific community. Meadowcroft Rockshelter has been awarded recognition by both the state and federal governments as having significance as a First American excavation site.

Pennsylvania Historical and Museum Commission and National Park Service plaques greet visitors as they enter Meadowcroft Rockshelter as documenting the presence of humans in North America "nearly 16,000 years" and "at least 16,000 years" ago, respectively.[19] These were statements by governmental organizations with recognized scholarly expertise—as well as political and funding power—declaring that not only must the Clovis-first model be incorrect, but also that Meadowcroft Rockshelter is to be accepted as having a special status in the search for the First Americans because of its extreme antiquity. Although the significance of simple historical markers and the motivations behind their being erected might be questioned, governmental power plays an influential role in how the search for the First Americans is conducted. Today, the Heinz History Center, which funds the project, states on its website that the Rockshelter "features 19,000-year-old evidence of the region's earliest inhabitants."[20]

Adovasio has also succeeded in making the excavation site itself easily accessible to the public at large, and also in preserving his archaeological methodology for inspection by both the public as well as First American scientists. The Rockshelter has been attractively covered against the elements, and the locations of the items of material evidence removed for analysis have been meticulously tagged. The modern American commoditization of First American heritage is captured at Meadowcroft Rockshelter where, for a fee, daily tours are available to the public accompanied by a knowledgeable, but amateur, guide.

# CHAPTER 5

# Case Studies: Atlantis and Chinese Bestiary

I N THE INTRODUCTION, myth was defined as being a statement concerning the natural world that is widely held within a culture and is generally accorded a status of being true, but at the same time is not necessarily (but might be) demonstrable according to the scientific methods of the time. Myth will continue to be used in that manner here to describe theories that are not demonstrable according to the science of our time but may still be believed as true in some quarters. Some may call such discredited notions "pseudoscience," which means not authentic science despite it possibly appearing science-like. However, once such concepts have been debunked by the scientific method itself, there is no "pseudo" involved; they are simply not science. Racial science no longer even has the trappings of science. To call such theories pseudoscience would inappropriately award them an unmerited status not available to American Indian creation myths. Myths reflect that they are believed by some elements of society without having a confirmed scientific basis. Distinguishing between a scientific theory or a traditional myth becomes easier with the luxury of employing a historical context. Some theories that are held to be scientific in one generation are believed to be mythical, or even falsified, by the next. The first case study examines the evolution of Atlantis from a nineteenth-century science to a twentieth-century myth as a source for the First Americans. The second investigates Chinese mythology as a potential historical source for myths of an original Asian homeland for the First Americans.

## Atlantis

One of the more conceptually elastic theories is that the First Americans came from Atlantis. Such flexibility allowed the indigenous origins to intertwine Plato, the bible, geography, anatomy, geology, linguistics, and material artifacts. Genetics and carbon dating were among the few sciences missing from the late nineteenth-century search for the First Americans. The ontological beauty of offering

Atlantis as an explanation is that a detractor—in an Alice-in-Wonderland application of Popper's falsifiability postulate—must demonstrate that something that then did not exist, indeed, never existed. How does one scientifically falsify the creative power of the human imagination?

Those that espoused the Atlantean theory have included some well-respected intellects employing what appeared to be a scientific methodology accompanied by substantive evidence. As Everett Franklin Bleiler (1820–1910)—a scholar of science fiction and pseudoscience—concluded, not only are the originators of enduring myths frequently unknown, a single person is rarely credited with the creation of more than one noteworthy myth. However, Ignatius Donnelly not only has the distinction of having created three of what Bleiler terms "golden" myths, but also did so believing, at the time, that he was carrying out modern science. In addition to the myth of Atlantis as a source of the First Americans, Bleiler credits Donnelly with espousing not only a mathematical proof that Bacon had written Shakespeare's plays and hidden a cipher message in them, but also that geological features are the result of catastrophes from the sky.[1] Typical of most enduring myths, there is enough of a factual basis in each of these to sustain it in a suggestive imagination.

Although popular belief might contend that one role of science is to debunk myth, Donnelly's work is interesting as a demonstration that the practice of science itself can be a source of what may one day turn out to be myth widely considered to be of the fantastical type. Of particular interest in the search for the First Americans is that in *Atlantis: The Antediluvian World*, which Bleiler calls "the most influential pseudo-scientific work of the later nineteenth century,"[2] Donnelly laid the foundation for the belief that Atlantis was the original birthplace of the American Indian. This work is cited by modern archaeologists as a (formerly) scientifically based statement that "the existence of the mid-Atlantic continent of Atlantis was not fable but historical fact."[3]

Plato's *Timaeus* and *Critias* have generally been credited as providing the first instance of the existence of Atlantis. Donnelly himself declared that there is nothing improbable in Plato's narrative. "It is a plain and reasonable history of a people."[4] In searching for a factual basis for Plato's account, scholars over the last two centuries have attempted to draw parallels between Atlantis and ancient civilizations in Minoan Crete, Egypt, and Persia. In an odd parallel to American Indian myths, Donnelly (among others) believed that Plato's account could be taken literally, and that there was a matching historical reality.[5] For Donnelly, Atlantis was not a myth, but rather had a substantive association with history. Despite the now-seemingly ludicrous nature of the idea of an Atlantean colonization of the New World, Donnelly's mid-nineteenth-century theory did not conflict with the little hard information on pre-Columbian civilizations available at that time.[6]

Donnelly's *Atlantis* had much the same appearance typical of other early ethnographic books of the mid-nineteenth century. There are charts, graphs, maps, portraits, alphabetic tables, and architectural drawings. Donnelly believed that the pre-Columbian civilizations of the Americas were the result of colonies launched from Atlantis and were based on the Atlantean culture and language.[7] He believed that the "shores of Atlantis were not far distant from the West India Islands; a people possessed of ships could readily pass from island to island until they reached the continent.... [A]ll the traditions of Central America and Mexico point to some country in the East, and beyond the sea, as the source of their first civilized people; and this region, known among them as 'Aztlan.'"[8]

Although the end result may have been inadvertently to create an enduring myth, Donnelly is credited with using facts, theories, and methodologies that were generally acceptable in his time. His analysis of Atlantis "had a certain low degree of fit within the knowledge of 1880. By 1890 it had far less, and by 1900 it had none at all."[9] Bleiler concluded that Atlantis cannot now be considered serious science and was inaccurate in its major conclusions.[10] Yet, Donnelly's analysis was predicated upon the standards of then-modern science, and contained many of the elements today considered to be scientifically acceptable methodologies.

Donnelly's first task was to physically position Atlantis to support his theory. If Atlantis were to have existed, he placed it along the Atlantic Ridge, a geologically appropriate location, especially considering that plate tectonics was not accepted by the geological community until seventy-five years later.[11] Donnelly also included in his analysis a search for commonality in material artifacts, a practice still central to the scholarship of current First American science. He attributed a common Atlantean cultural source as explaining the similarity in arch construction in Mycenae and Central America. In what might have been an early expression of a central conceptual flaw at the root of social constructivists, it apparently never occurred to Donnelly that the similar approaches to arch construction were attributable to the universal constancy of gravity and compressive/tensile strength of stone, rather than being attributable to a common socially constructed approach to arch building in the two locations.[12]

As is the case in modern First American science, Donnelly also employed, albeit very rudimentarily, the tools of bioanthropology, specifically craniometry, to support his case for Atlantis. He posited that there was a highly similar cranial shape between that of the indigenous Central Americans and that of ancient Egyptians. This commonality, he concluded, was a result of their common ancestry in Atlantis.[13]

Again, prefiguring the approaches used in the current search for the First Americans, Donnelly conducted a detailed linguistic analysis in an attempt to find commonality between modern languages that could be used to support Atlantis as the source for populating the New World. Not surprisingly, Donnelly contended that he had found just such evidence. As James Adovasio characterized it, Donnelly found the linguistic similarity between the words Aztlán and Atlantis too obvious to be coincidental.[14] Thus, as was the case with the construction of arches, Donnelly had his connection. He believed he had discovered the reputed linguistic commonality between the Dakota Sioux dialect and other languages, including Latin, English, and Sanscrit (*sic*).[15] He also analyzed alphabets—including Mayan, Irish, Hebrew, and Ethiopian—and again found a basis for Atlantis as a First American source.[16]

Donnelly exhibited some nimble reasoning in tying together Asiatic migration, Beringia, Atlantis, and assorted catastrophic floods. "It has been the custom to ascribe the recognized similarities between the Indians of America and the Chinese and Japanese to a migration by way of Behring's (*sic*) Strait from Asia into America; but when we find . . . such distinct recollections of the destruction of Atlantis in the Flood legends of the American races, it seems more reasonable to conclude that the resemblances between the Othomi [a Mexican "race"] and the Chinese are to be accounted for by intercourse through Atlantis."[17]

Ignatius Donnelly was not the only scholar in the early development of First American science who addressed the role of Atlantis. In *Prehistoric Races of the United States of America*, in 1873 John Wells Foster, then-president of the Chicago Academy of Sciences, also situated Atlantis as a central part of his thesis for explaining the populating of the New World. Citing the work of Charles-Étienne, Abbé Brasseur de Bourbourg (1814–74), a French archaeologist and ethnographer, to whom he attributed the beginnings of American ethnology, Foster agreed that "the words Atlas and Atlantic have no satisfactory etymology in any language known to Europe. They are not Greek and cannot be referred to any known language of the Old World. But in the Nahuatal (or Toltecan) language we find immediately the radical a, atl, which signifies water, man, and the top of the head."[18] Foster did, however, discount assertions by George Catlin (1796–1872) that such ancient cities as Palenque and Uxmal are tied to Atlantis because Catlin contended that the evidence showed that "the ocean has been their bed for thousands of years, and that the earth on which one treads, and the whole face of the country in which they stand, bear incontestable proofs of the same fact."[19] It may seem ironic to us that Foster, a leading scientist of the nineteenth century, could accept what now appears to be such specious reasoning with regard to linguistic similarities, yet dismissed an argument that Palenque or Uxmal might have descended from an Atlantean civilization because the geology appeared oceanic, a theory that modern plate tectonic theory might now support.

As a case study, Atlantis demonstrates the full range of issues associated with the search for the First Americans. First, it provides an example of the difficulty, except in retrospect, of distinguishing science from myth. Second, in the latter half of the nineteenth century, reputable scientists could make a case based on acceptable data and methodologies that not only did Atlantis exist, but that it was also a prime candidate for the origin of the First Americans. This, of course, is now considered both dubious science and a specious conclusion. The sciences evolve; what is considered to be acceptable in one generation may well be considered erroneous in the next. Finally, it provides an example as to how myths can endure by co-opting the trappings of scientific authority.

## Chinese Bestiary

The focus of the search for the first Americans is physically as well as conceptually centered in the Western Hemisphere. It is an investigation into where people came from—Atlantis, Solutrean-Europe, Asia via Beringia, or Phoenicia. Since many First American scientists believe Beringia was the path for the first migration, looking to Siberia for precursor cultures might prove illuminating. However, as David Meltzer acknowledged, "Siberia is so poorly known archaeologically that there is no reason to believe that the earliest sites there have yet been found. . . . Before we can talk about the peopling of Siberia, Siberia will have to be peopled by archaeologists."[20]

Outside of the search for genetic or linguistic roots, there is not a substantial core of scholarship that investigates the cultural myths that might lend some insight into Asian people that went to the New World. If American Indians are correct that their myths retain an accurate representation of their existential experience of thousands of years ago, a case might be made that the myths of the possible mother ship cultures may have some bearing on the First Americans; there might be a cultural myth of having sent people to the New World.

The paucity of scholarship is, of course, principally a function of the improbability of examining any artifacts of the Atlantean, Phoenician, or Solutrean cultures; examination of written records is an impossibility. There is, however, one viable option in the written records of China. The Chinese have the longest continuous record of written history on Earth. While the genetic evidence indicates that modern American Indians are not direct descendants of the Chinese, but rather are descendants of a common ancestor, ancient Chinese records have frequently been consulted by Western scholars as a scientific source of confirmation for otherwise mythic events such as floods and celestial occurrences. Bones and turtle shells with written histories date approximately to 1200 BCE during the Shang Dynasty (c. 1700–1046 BCE). This indicates not only an early and sophisticated system capable of preserving a written system of cultural beliefs, but also that three thousand years ago the Shang culture already had an ancient history that they believed was worth preserving. If the American Indians have an oral tradition of their existential origins that is difficult to document beyond a few generations, but they would have anthropologists believe reaches back thousands of years, the question arises as to what written records might exist in the Chinese culture that would indicate a knowledge that any people, Chinese or otherwise, might have departed toward the northeast of Asia, especially in view of the geneticists' hypothesizing that, after the initial migration from Asia to the New World via Beringia, there was doubtless a two-way migration of both humans and animals.

In 6 BCE, the Western Han dynasty Emperor Ming was presented with a completed compendium that was a several-hundred-year effort to capture knowledge that Chinese scholars believed was being lost. The *Guideways through Mountains and Seas* (山海经 *Shānhǎi Jīng*), written between the fourth and first centuries BCE, is perhaps the classic ancient history of Chinese beliefs. According to a modern translation by Richard Strassberg, it "is a unique and enduring record of a wide range of beliefs held by the ancient Chinese about their world, encompassing religion, mythology, geography, flora, fauna, minerals, and medicine." The geographical import of the work is evident in that, while 经 (jīng) may be translated as guideways, classic book, or collection, 山 (shān) and 海 (hǎi) are geographical terms meaning mountain(s) and sea(s), respectively. Strassberg calls it an "encyclopedic cosmography."[21] It is an authoritative cultural statement. As best understood by the Chinese of 2,400 years ago, it was ancient history, not myth.

Unlike American Indian myths, which were exclusively oral in nature, the Chinese were not only among the first to use written symbols, but they also then proceeded to make, and retain, extensive records of the monumental as well as the mundane. *Shānhǎi Jīng* is a catalogue of the natural world within and surrounding the ancient Chinese homeland. The first part can be considered a map of sorts showing the location of particular mountains, plants, and minerals.[22] However, as Strassberg indicated, *Shānhăi Jīng* also "identifies some ninety-five foreign lands and tribes.... The ethnographic data includes peoples confirmed in historical texts as well as highly fantastic groups located in mythological realms."[23] The more interesting foreign people are endowed with ears so long that they must use their hands to carry them, and others are variously one-eyed, one-armed, or without intestines.[24]

*Shānhăi Jīng* described China as being surrounded not only by shān, mountains, but also by hăi, seas, that are believed to be both physical oceans and, allegorically, the vast expanses of desert on China's western borders. Chinese scholar Zhang Huang (章潢) (1527–1608) attempted in *Compendium of Diagrams* (图书编, *Túshūbiān*), published in 1613, to capture pictorially the knowledge that had been amassed in ancient China, including that of the *Shānhăi Jīng*. According to Zhang, the existential myth of China was perceived to be of a culture surrounded by vast expanses of deserts and seas. To the northeast of the diagram were the Long Thighs People and the Long Arms People; to the southeast was the Land of the Midgets; to the southwest, the Land of Gentlemen; and to the northwest, the Land of the People without Calves.[25]

Interestingly, there are lands identified beyond the seas with non-Chinese inhabitants, including to the northeast, which would be the direction of Beringia. There is, however, no indication of a migration by the Chinese or by the foreigners of the northeast that might support the Beringian hypothesis. Nor is there any indication of a two-way migration between Asia and the New World as suggested by geneticists.

The absence in Chinese mythology of any indication of a migration, either by sea or across a land bridge, to the New World cannot be considered disproof of either those American Indian oral histories that speak of an existential creation or of modern science that proposes a migration from Asia. It does, perhaps, suggest the unimportance of the peopling of the Western Hemisphere in the myths of the Old World.

# PART II

# First American Science

# CHAPTER 6

# Anthropology and Archaeology

T HE PRACTICE OF FIRST AMERICAN SCIENCE is a problematic undertaking given the existential challenges to archaeology and anthropology from inside as well as outside the disciplines. Leading anthropologists seriously proposed that anthropology is not a science; archaeologists are uncomfortable as a subdomain of anthropology; archaeologists also question whether their discipline is a science; and both anthropology and archaeology must survive in a postmodernist world that has shifted the ontological balance between American Indian myths and First American science.

The internecine squabbling within science can seem outright bizarre to those outside of the community involved. For example, in 2006, the International Union for Quaternary Research (IUQR) wrote to the International Commission on Stratigraphy (ICS) to complain that the ICS had defined the Pleistocene epoch in such a manner as to cover only a part of the ice age. The IUQR demanded that if the ICS did not acceptably redefine the Pleistocene, then the IUQR would consider it a "unilateral and hostile" act.[1] As David Meltzer commented, "we demanded nothing less than control over our period of geological time."[2]

Archaeology rests uneasily within the intellectual embrace of anthropology. In the United States, it has been relegated to a subdiscipline of anthropology since the time of Franz Boas (1858–1942), one of the founders of American anthropology.[3] The dynamics across archaeology, anthropology, and science in general are complex and contentious. In "Why Don't We Know When the First People Came to North America?," Meltzer stated that "demonstrating that there were humans in America before 12,000 BP is strictly an archaeological issue."[4] This is not just a defense against American Indian myths but also an attempt to erect a barrier against the potentially more powerful intellectual domains presented by other sciences. It is not that Meltzer, whose writings would lead one to consider him a statesman rather than a radical in archaeological politics, does not recognize that other sciences might provide some insight. "Data from an array of fields would seem to narrow the number and timing of migrations [from Asia to the New World], but that evidence is at best circumstantial and cannot be used to constrain what is strictly an archaeological matter."[5] It appears that data can be tainted by not being produced archaeologically. Archaeologists are certainly not oblivious to the contributions of the other First American sciences, but they apparently believe that they, as archaeologists, should dictate what results from the other sciences are to be considered legitimate in the search for the First Americans.

There are, however, contrary opinions from within the discipline. As Randall McGuire has indicated, archaeologists should respect the power inherent in their community while recognizing the politics involved. As McGuire stated, "Archaeology is always political . . . but this does not mean that we should give up our authority as good crafts persons. Speaking truth to power requires that we maintain the authority of our craft."[6]

There has not always been an acknowledgment of the political aspects of science. Thomas Jefferson, one of the first American practitioners of archaeology, in 1809 maintained a belief in the conflict-free nature of science. "These [scientific] societies are always in peace, however their nations may be at war. Like the republic of letters, they form a great fraternity spreading over the whole earth, and their correspondence is never interrupted by any civilized nation."[7]

Conventional wisdom would have us believe that the soft sciences—such as sociology and anthropology—covet the greater cultural authority Western society grants to the hard sciences—such as chemistry and, especially, physics. Archaeology is stigmatized, in some quarters, by those who consider it to be "a social science—which means almost automatically soft and imprecise."[8] These soft sciences, so this wisdom goes, attempt whenever possible to mimic the methodologies and vocabularies of the hard sciences.

Those modern intellectual disciplines that profess to most accurately characterize human interactions with the world are typically self-described as being based on science—unless, apparently, that discipline is anthropology. Richard Feynman, one of the preeminent physicists of the twentieth century, is reported to have stated that "philosophy of science is about as useful to scientists as ornithology is to birds."[9] Interestingly, anthropology, the epistemological approach central to the search for the First Americans, was recently embroiled in a controversy over distancing itself from science and engaged in a philosophical discussion as to what constitutes a meaningful way of knowing nature.

The American Anthropological Association (the AAA) was founded in 1902 and has advertised itself as being "the world's largest association for professional anthropologists, with more than 10,000 members . . . and covers all four main fields of anthropology (cultural anthropology, biological/physical anthropology, archaeology, and linguistic anthropology)."[10] On November 20, 2010, an existential crisis in anthropology erupted when the Executive Board of the AAA adopted a new long-range plan that would have removed anthropology as a scientific endeavor. As contained in an openly broadcast email to anthropologists and the public from Peter N. Peregrine, president of the rival Society for Anthropological Sciences, the edited text of the proposed controversial changes was as follows (the additions are underlined, and the deletions are struck through):

> Section 1. The purposes of the Association shall be to advance ~~anthropology as the science that studies~~ <u>public understanding of</u> humankind in all its aspects~~., through~~ <u>This includes, but is not limited to,</u> archeological, biological, ~~ethnological,~~ <u>social, cultural, economic, political, historical, medical, visual,</u> and linguistic <u>anthropological</u> research; <u>the Association also commits itself</u> ~~and~~ to further the professional interests of ~~American~~ anthropologists, including the dissemination of anthropological knowledge, <u>expertise, and interpretation.</u> ~~and its use to solve human problems.~~
>
> Section 2. To advance ~~the science of anthropology~~ <u>the public understanding of humankind,</u> the Association shall: ~~Foster and support the development of special anthropological societies organized on a regional or functional basis;~~ Publish and promote the publication of anthropological monographs and journals; Encourage anthropological ~~teaching,~~ research <u>and practice;</u> ~~act to coordinate activities of members of the Association with those of other organizations concerned with anthropology;~~ and maintain effective liaison with related ~~sciences~~ <u>knowledge disciplines</u> and their organizations.
>
> Section 3. To further the professional interests of anthropologists, the Association shall~~, in addition to those activities described under Section 2. Take action on behalf of the entire profession and integrate the professional activities of anthropologists in the special aspects of the science; and~~ promote the widespread recognition and constant improvement of professional standards in anthropology.[11]

The AAA executive board had deleted all specific references to science as a part of anthropology. In addition, it had deleted the reference to "American" in section 1. The mission was no longer to "advance the science of anthropology" but rather to "advance the public understanding of humankind." Geoffrey A. Clark, at Arizona State University, concluded that the AAA over the past couple of decades had been undergoing an intellectual Balkanization, that anthropology had no discernible conceptual framework, and that anthropology was "divided against itself (and therefore rather incoherent)."[12] Anthropology was born through the separation of science from myth. Apparently, the leaders of the AAA were on a path to their reengagement.

Reflecting the popular status of anthropology in American culture, these changes were widely addressed in press articles ranging from the *New York Times* to *Psychology Today* and *Inside Higher Ed*. The *Times* captured the crux of the issue: "The decision has reopened a long-simmering tension between researchers in science-based anthropological disciplines—including archaeologists, physical anthropologists and some cultural anthropologists—and members of the profession who study race, ethnicity and gender and see themselves as advocates for native peoples or human rights."[13] In an email to the *Times*, Virginia Dominguez, then president of the AAA and a University of Illinois anthropologist, indicated that the word science "had been dropped because the board sought to include anthropologists who do not locate their work with the sciences."[14] Although it may be appropriate for the organizational documents of a professional association to reflect whatever membership they desire to recruit and whatever professional standards they hope to maintain, it is, however, difficult to envision physicists or chemists situating the conceptual center of their preeminent professional associations outside of science as anthropology attempted to do.

In a November 25, 2010, article in *Psychology Today* titled "No Science, Please, We're Anthropologists," Alice Dreger—who earned a PhD in the history and philosophy of science from Indiana University and is a former professor of clinical medical humanities and bioethics at Northwestern University's Feinberg School of Medicine—described those who wish to separate anthropology from science as "fluff-head cultural anthropological types who think science is just another way of knowing. . . . You can usually tell the ones who are fluff-heads by their constant need to look like superheroes for oppressed peoples, and you can tell the non-fluff-heads by their attention to data."[15] In an email to Dreger, Stuart Plattner, formerly program director for cultural anthropology at the National Science Foundation, called the revised AAA platform "another step in the conversion of Anthropology from a social science into an esoteric branch of journalism."[16] Raymond Hames, chair of the Anthropology Department at the University of Nebraska–Lincoln, stated that "advocacy is what we do as citizens in a democratic society. Even as anthropologists we must advocate on the basis of fundamental science. Science has a special currency in courts, public opinion, and in the legislative process. If we purge science from our mission statement we lose our credibility, the ability to advocate for effective change, and hence our power to do good. We become just another special interest group."[17] In a November 30, 2010, *Inside Higher Ed* article titled "Anthropology Without Science," Hames also asked that if children were dying of dysentery should scientists accept as an explanation that a violated taboo has made the gods angry, or should they determine how fecal matter might have contaminated the water supply?[18] The *Inside Higher Ed* article described the debate as: "to one group, objective truth as revealed by science is an ideal to pursue, while to the other, that notion poses problems because it embodies Westernized and colonial ideals. . . . Some argued that being an anthropologist, by necessity, meant that one had to advocate on behalf of one's subjects."[19]

Just as observations are theory-laden, theories that attempt to describe ways in which the natural world functions are themselves culture-laden. Any methodical approach structures the subject matter such that certain kinds of explanation are appropriate, and others are not.[20] This does not, however, mean that all approaches are equally effective in providing functionally useful characterizations of the natural world. Questions can be framed only if potential answers can be imagined, and any particular culture stimulates and simultaneously constrains such imaginings. It appeared that the AAA was attempting to position its epistemological domain outside of the American cultural operating assumptions, whatever they might be, that govern both science and traditional knowledge. Even if the AAA could succeed in finding that special epistemological place, it is not clear what, if anything, could be accomplished through such an approach other than better positioning a politically correct AAA in their envisaged postmodernist world. Whether science does or does not better capture the operational characteristics of the natural world is less important for the success of the AAA's effort than the fact that practitioners of science believe that to be the case and may seek membership in a professional association that is compatible with that view. The distinction to be made here is between what constitutes a culturally appropriate approach to nature—the particular methodologies of science or the vast range of traditional lore—as opposed to how scientists themselves structure the professional associations of their discipline. The AAA leadership indicated that they believed that, however the profession of anthropology might be defined, at a minimum it does not require a commitment to science; many approaches to nature would be welcome. According to Edward Said, "of all the modern sciences, anthropology is the one historically most closely tied to colonialism, since it was often the case that anthropologists and ethnologists advised colonial rulers on the manners and mores of native people."[21] Claude Levi-Strauss called anthropology "the handmaiden of colonialism."[22] The AAA was attempting to situate itself as outside of the shadow of such accusations.

The membership revolt successfully halted the rejection of science by the AAA leadership. Currently, the AAA Statement of Purpose is:

> Section 1. To advance anthropology as the science that studies humankind in all its aspects, through archeological, biological, ethnological, and linguistic research; and to further the professional interests of American anthropologists, including the dissemination of anthropological knowledge and its use to solve human problems.
>
> Section 2. To advance the science of anthropology, the Association shall: Foster and support the development of special anthropological societies organized on a regional or functional basis; Publish and promote the publication of anthropological monographs and journals; Encourage anthropological research, act to coordinate activities of members of the Association with those of other organizations concerned with anthropology, and maintain effective liaison with related sciences and their organizations.[23]

The central questions in the search for the First Americans are straightforward: where did they come from; who were they; when did they come; are any of their descendants still living today? As Meltzer pointed out, "[T]here is only the broadest agreement among the different disciplines on the central questions related to the peopling of the Americas."[24] As can be seen in the debate over the AAA mission statement changes, difficulties arise in the fundamentally dissimilar approaches that the different sciences take in translating these seemingly simple questions into theoretical structures that shape their research regimes and, ultimately, their observational boundaries. At times, it even appears that a theory purporting to explain the arrival of the First Americans can assume some of the cultural trappings normally associated with a mythological status. According to the archaeologist Tom Dillehay, the power of the Clovis-first paradigm as an explanation for the First Americans effectively excluded some paths of research, and its demise has produced "fresh thinking."[25] James Adovasio called it the "Clovis Bar."[26] Moreover, reputable archaeologists appear to believe that findings from the nonarchaeological sciences must fit into archaeological practice. Meltzer stated that it is "wrong to dismiss archaeological evidence because it fails to match the chronological expectations derived from nonarchaeological sources. . . . If a site is found that is older than it 'should' be based on evidence from language, teeth, and genetics, then that nonarchaeological evidence will have to be adjusted accordingly."[27] In instances such as this, it is difficult to distinguish between a religious practice that denies legitimacy to material evidence challenging a revered myth, and the manner in which First American science can be practiced by denying legitimacy to material evidence that challenges a reigning archaeological paradigm. The public, Vine Deloria Jr. has said, needs to know how "does archaeological speculation differ from Erich von Daniken's citation of the Nazca lines as evidence of early spaceman?"[28] Although interdisciplinary approaches may prove productive, as Meltzer feared, they may also generate theoretical contradictions, methodological conflicts, and *ad hoc posterior* reinterpretations. "Neither genes nor languages can be dated; only archaeological materials can," wrote Meltzer.[29] Yet the dating mechanism that archaeological materials rely on is borrowed from physics. Furthermore, the carbon-14 decay that forms the basis of dating First American-associated artifacts, must be translated from radiocarbon years to chronological years based on counting tree rings or ice cores for the past fifteen thousand years. While First American scientists have grown accustomed to this process, it should be noted that the archaeological dates given are expressed in terms of the probability of the date being included within a specified range. Given the nature of the debate, there is more at stake than simply providing chronological markers. The inter-domain boundaries are quickly drawn. The issue is how the different disciplines interact and how that interaction leads, if it does, to a clearer picture of the First Americans.

Critiques of the rigidity associated with archaeological theories come from both inside as well as outside the profession. Keith W. Kintigh, a professor of anthropology at Arizona State University, commented that a considerable portion of archaeology is still "theoretically in an earlier age."[30] Deloria stated, "In methodological terms there is a major problem in bringing non-Western traditions within the scope of serious scientific perspective, and that is the inherent racism in academia and in scientific circles. Some of the racism is doctrinaire and unforgiving—for instance, the belief that, for a person or community possessing any knowledge that is not white/Western in origin, verification and articulation are unreliable."[31] Despite the hurdles suggested by Deloria, the archaeologist Thomas W. Killion suggested that the practice of archaeology has begun to change as a result of the challenges presented by the non-Western traditions.[32] Although the self-image of First American archeology presents itself as if it were a positivist science, it is far from clear whether its knowledge is truly cumulative or simply a succession of new "facts." Geoffrey A. Clark, Regent's Professor Emeritus in the School of Human Evolution and Social Change at Arizona State University, is concerned over "the fad-like, insubstantial nature of what passes for 'theory' in American archaeology. . . . Hypotheses are never really confirmed or disconfirmed (most of them cannot even be formulated in a testable way), questions are never really answered in archaeology—after a while, they simply ceased to be 'interesting.'"[33]

Even as these fundamental tensions ripple across anthropology, archaeology, and American Indian myths, there is also an existential turmoil internal to archaeology itself. The first significant concern to archaeologists is deciding whether they indeed are scientists. According to David Killick and Paul Goldberg, even when university archaeology programs offer coursework in what is termed archaeological science, graduate students are not required to take it, and most do not.[34] As a result, "most archaeologists don't know enough science to be able to collaborate effectively with scientists who don't know any archaeology."[35] As a part of a series of dueling articles, Geoffrey A. Clark expressed concern over the "contention that archaeology is a science in its own right."[36] Michael Shott expressed doubt as to whether archaeology offers any explanatory power since it only describes the material record and then borrows anthropological theory to devise simply "plausible accounts" of the past.[37]

Whether archaeology is or is not a science, there also appears to be no clear consensus as to whether archaeology is, or even should be, practiced independently of anthropology. Killick and Goldberg insisted that archaeology is "a discipline in its own right."[38] Clark countered that archaeology is not a discipline separate from anthropology.[39] As a result of all of these perturbations, Shott has eloquently concluded that "the archaeological patient is ill."[40] James Adovasio, in his typically blunt manner, believed that the contentiousness internal to First American science has not only produced "an astounding amount of silliness and even profound stupidity has been taken as serious thought," but also raised the question as to "whether the field of archaeology can ever be pursued as a science."[41]

CHAPTER 7

# Bioanthropology

A POPULAR AMERICAN TELEVISION MOTIF is that of an esoteric specialist who is called in to examine human remains as a part of a police investigation. On television, this expert might calculate certain skeletal ratios, examine the shape of the skull and pelvis, and study the dental structure and wear patterns. Based upon the results, the expert extracts a range of information from the remains—as well as the environs within which they are discovered—including sex, race, age, time since death, and incidents of major skeletal trauma possibly including cause of death. As James Chatters, a principal expert involved in the Kennewick Man case, characterized it, in the context of a crime scene such an expert is practicing forensic anthropology; when the human remains are ancient and are not related to a criminal investigation, the specialist is practicing bioanthropology.[1] The same trained investigator using identical techniques on two anatomically Homo sapiens skeletons can be considered to be practicing different sciences depending upon the final consumer of the data.

Although genetics now assumes center stage in the application of the biological sciences to the First American conundrum, the study of ancient human skulls and teeth have both made substantive contributions to the debate over the last century. Tension between genetics and bioanthropology for primacy of scientific place is recurrent in the academic literature. For example, a group of First American scientists stated that "the cranial morphology of the earliest Americans . . . is significantly different from that of more recent Native Americans. . . . Accordingly, Paleoamericans came to the New World first and were later replaced by ancestors of modern Native Americans. Genetic data do not support this model. . . . Thus, although the Paleoamerican sample is still small, the morphological differences are likely the result of genetic drift and natural selection, not separate migrations."[2] Here we have an authoritative statement that, at once, cites the results of craniometry, yet simultaneously awards a privileged position to the results of genetics. The study of teeth also provides both illumination and contradiction for the results of the other sciences. The academic jockeying is evident in the writings of scientists themselves. For example, Stuart Fiedel believes that "the most convincing evidence for the Asian origin of the Native Americans comes not from archaeology but from physical anthropology [i.e., bioanthropology] and geology."[3]

Science cannot be practiced in the absence of data. One of the constraining factors for bioanthropologists involved in the search for the First Americans is the limited number of temporally suitable human remains available for examination. There are only a few dozen ancient skeletal remains that have ever been discovered in the United States, few of them are complete, and an increasing number have been claimed for reburial by American Indian groups and are therefore no longer available for analysis.[4]

Racism, in fact or as a diversionary debating stratagem, looms large in the search for the First Americans. For First American scientists, "the concept of race is best described as an expression of cultural ideology and not a biological reality."[5] Although modern archaeological theories must be sensitive to the racist history of Western archaeology, it is not entirely clear when such sensitivity crosses a boundary from being concerned with objectivity and becomes a political correctness that attempts to enforce a particular mindset. Sociologists and geneticists have largely disavowed contentions that there is a meaningful difference (i.e., distinctions beyond phenotype) between races; that assumes, of course, that race itself could be sufficiently well defined for there to be a metric created to measure it. In 1996, the American Association of Physical Anthropology described the concept of pure races as being "untenable."[6]

Biological morphology is the study of the structures and forms of organisms. Craniometry is a subfield of morphology dealing with the measurements of human skulls. With a consensus that morphology is a product of genetic determination, humans related through a closely linked ancestry are believed to share similar physical features and, therefore, craniofacial dimensions.[7] The results of craniometrical analysis have been employed since the nineteenth century by social theorists to coopt the authority of science in an attempt to justify racial stratification. Many of the same arguments that have been employed in the race-as-a-metric debate have been employed in the debates over the significance, if any, of differences in skull shape and size.

The application of the newly emerging methods of nineteenth-century First American science was paramount in the attempts to document the viability of a relationship between skull characteristics and a racial differentiation in capabilities. By the early twentieth century, however, research findings of the likes of Franz Boas challenged the foundation of craniometry. In 1911, Boas "demonstrated that the skulls of Jewish and Italian youths born in New York differed markedly from those of their immigrant parents. . . . Furthermore, Boas showed that head and body form varied more widely within than between populations, and that there was often considerable overlap in the measurements of different races."[8] Despite such findings, in some sectors there continued to be a belief that science not only could, but in fact had, documented a substantive association between biologically based human factors such as intellect, on the one hand, and phenotypically expressed traits such as skin/hair color and skull size, on the other. According to Fiedel, First American science has, at times, reflected "the racist assumption that Native Americans were backward savages, incapable of devising sophisticated cultures without the benevolent assistance of more advance white-skinned tutors."[9] Perhaps the apex of such pseudoscientific associations was evident in the Nazi use of physical metrics of racial classification to exterminate Jews and the Roma during World War II.

A belief that modern American Indians were incapable of originating the cultures found in the New World led to what might be better termed as speculations rather than theories. For example, the rise of the Olmec civilization has been attributed to Phoenician exploration of Mexico's east coast as evidenced by a bearded, Semitic face, called "Uncle Sam," on a stela at La Venta. The Olmec utilization of jade and the religious use of cat-like figures have been attributed to China's Shang dynasty. Finally, some have speculated that the use in Mesoamerican art and architecture of people sitting cross-legged or the depiction of water lilies can be credited to Buddhist missionaries.[10]

Craniometry has been a long-standing part of the scientific tool kit in the search for the First Americans. In the late nineteenth century, John Wells Foster included in the *Pre-Historic Races* a thirty-six-page chapter on the "Crania of the Mound-Builders."[11] While acknowledging that the cranial variations within any particular race are broad, Foster concluded that "in a large assemblage of skulls derived from a particular race, there is a general conformation, a predominant type . . . which has been regarded among the surest guides in tracing national affinities."[12] Foster concluded that "my observations have led me to infer that the Mound-builders' crania were characterized by a general conformation of parts, which clearly separated them from the existing races of man, and particularly from the Indians of North America."[13] In other words, modern American Indians did not build the mounds. In describing skulls collected from midwestern American mounds, Foster also concluded that "many of these characteristics, which are not conspicuous in a well-developed European skull, indicate an approach towards the lower animals of the anthropoid type."[14] Foster's analysis using then-accepted craniometrical methodologies concluded that not only were Europeans scientifically proven to be the most advanced of the races, but that indigenous Australians were the least advanced, with the American Indian mound-builders falling somewhere between. He even created a schematic for a hierarchy of races based on skull size with the European skull demonstrating the highest development.[15]

In another result of his craniometrical analysis on the remains of an American Indian mound-builder, Foster stated that "No one, I think, can view this fragment of a skull, with the superciliary ridges projecting far beyond the general contour, both laterally and in front, and the low, flat forehead, with its thick, bony walls, without coming to the conclusion that its possessor was a ferocious brute. The prize-fighter of this day might envy such a frontispiece, adapted to withstand any amount of pommeling, or almost even to turn a musket ball."[16] Based solely on the shape of the skull, Foster was comfortable in projecting behavioral characteristics—a "ferocious brute"—onto the inanimate bones.

A predisposition to believe in a difference in mental capabilities—correct or not—gave rise to theories to explain the difference. Physical variations (e.g., cranial shape and size) were employed as causal factors for the believed disparity in mental capabilities. A value hierarchy of cultural differentiation was then established to be explained by these physical differences. If Foster and others had produced their research today, it is clear that their conclusions would be considered racist.

Distinguishing between a scientist's personal views and how science is practiced is not always an easy task for scholars. For example, Samuel G. Morton (1799–1851), a physician and professor at the University of Pennsylvania, held views that today might be considered racist, but he also concluded that the modern American Indians were indeed related to the Ohio mound-builders. Unfortunately, as James Adovasio commented, "Morton's findings were easily ignored by believers in a separate, 'higher' race," a view that Morton may otherwise well have endorsed.[17] In 1981, Morton's scientific practice was the subject of a modern attack by the famed Harvard biologist Stephen Jay Gould (1941–2002). In *The Mismeasure of Man*, Gould offered an analysis in which he claimed that Morton had—consciously or otherwise—manipulated his data to fit his racist bias.[18] Based upon Gould's analysis, Morton today has become "a canonical example of scientific misconduct."[19] However, in an interesting turn of events, in June 2011, Jason E. Lewis and five other leading anthropologists published an examination of Gould's review of Morton's work in which they concluded that it was Gould himself who had manipulated his data to fit a preconceived notion of Morton's bias and that Morton "did not manipulate data.... In fact, the Morton case provides an example of how the scientific method can shield results from cultural biases."[20]

The debate continues today as to the extent to which skeletal traits provide substantive information "of a past individual's or group's cultural identity, independent of archaeological context."[21] In the nineteenth century, the methodologies of the likes of Foster, Ignatius Donnelly, and others were, for a time, considered within the scientific mainstream. Now, their work is considered to have been unduly influenced by the racism of that time. Morton's approach, ignored at the time, was first discounted and finally endorsed by modern scholars. An aphorism frequently attributed to nineteenth-century German Chancellor Otto von Bismarck states that people should not watch either laws or sausages being made because the process is disconcertingly messy. Perhaps the methodologies of First American bioanthropology should be added to that list.

As is evident from the work of Foster, Donnelly, and Morton, nineteenth-century scientists were comfortable in making *culture-wide* projections based on *biological* data from a relatively few skulls. With the advent of the twentieth century, theories associated with bioanthropological typology became more dependent on population analysis. Douglas W. Owsley, division head for physical anthropology at the Smithsonian Institution's National Museum of Natural History, concluded that this shift resulted from advances in genetics and biology that forced scientists to recognize not only the diversity across, but also within, population groups.[22] The ability for First American scientists to implement this shift in doctrine was greatly facilitated in the second half of the twentieth century by the increasing capabilities for data manipulation offered by computers.

Working at the US Armed Forces Human Identification Lab in Tokyo in the 1950s with human remains collected during World War II and the Korean War, Mildred Trotter at Washington University created an anatomical database that provided stature formulas for Mongoloids (including Asians and American Indians), whites, black Americans, and Mexican Americans. As useful, however, as such a database might be, the remains of the earliest Americans that have been discovered are from a time prior to the divergence of some of these racial groupings.[23] The use of modern anatomical standards for judging ancient remains is further compounded by the widely acknowledged reality that the average statistical difference between Negroid, Mongoloid, and Caucasoid as races is less than the range of differences within each group.

Modern skull measurements are called craniofacial morphometrics, a methodology developed during the 1970s by William W. Howells, an anthropologist at Harvard University. Subsequently, Richard Jantz and Steven Ousley developed a software called ForDisc, which uses an unidentified skull's metrics to provide a probability of its matching any particular human population.[24] ForDisc is currently available for download from the University of Tennessee Forensic Anthropology Center for $395.[25] ForDisc has been used by law enforcement agencies to identify bodies by narrowing the pool of candidates for matching dental verification and DNA tests.[26]

Automated matching regimes can be extremely valuable in providing matches for modern human remains where there is a robust set of statistics in the database. Matches, however, can only be estimated for population groups that are contained within the database. An obvious limitation with such software is that it can only match an unknown skull to the nearest group of remains in the database. Consequently, the atypical remains of an ancient First American candidate are automatically matched against the population in the database that it most closely resembles, notwithstanding that the remains may indeed not be from that population group.

The examination of skulls, skeletons, and teeth would seem to be such a straightforward, apolitical discipline that it is hard to imagine that the practice of bioanthropology could be significantly impacted by national policy decisions. However, as a result of the changing political landscape in the United States, including congressional enactment of the Native American Graves Protection and Repatriation Act (NAGPRA), from 1995 to 2005 there was a decline in the production of bioanthropological academic papers related to American Indians, because "Native Americans are increasingly controlling the terms of how their ancestors' remains are treated."[27] Moreover, American bioanthropologists are reportedly increasingly conducting their primary research projects in other countries to avoid such constraints.[28]

Bioanthropology presents an example of dynamics outside of science that can substantively influence how or where a given science is practiced—or even determine what a particular science is. With regard to the particular scientific puzzle of finding the First Americans, not only must bioanthropology function within the boundaries it has negotiated with the other sciences also addressing the same questions, but it also finds that how (as constrained by NAGPRA and other political agendas) and where (for those scientists who shift their research outside of the United States) it may be practiced is bounded by considerations traditionally thought to be outside of the scientific domain. According to anthropologist Ann M. Kakaliouras, bioanthropology is "no longer a monolithic scientific discipline."[29] Indeed, it is likely that it never was one, but it may be that bioanthropologists are just now coming to recognize that they have always functioned within a larger social context that structured their discipline.

The difficulty in entirely dismissing craniometry is that it provides scientists a way of seemingly associating human artifacts independent of cultures. Stuart Fiedel concluded that there is sufficient similarity between the mid-line ridge on the top of the skulls of Homo erectus discovered in the 1930s in Choukoutien, China, to "strongly suggest" a genetic contribution to modern Asians and American Indians.[30] Craniometrical analysis also led James Chatters to conclude that Kennewick Man, Stick Man, Horn Shelter Man, and Spirit Cave Man—all ancient human remains discovered in North America—had skull characteristics that were not in common with those of modern American Indians.[31] What is not clear, however, is the extent to which differentiating between skull shapes allows First American scientists to draw not only possible biological distinction, but even cultural ones as well.

In addition to craniometry, dentochronology is a separate and important subfield within bioanthropology that examines the morphology and wear patterns on teeth in order to temporally situate human remains as well as to draw ancestral relationships.[32] Although the examination of teeth as a tool in the search for the First Americans has been less prominent than that of the examination of skulls, it is not a recently developed methodology. In his nineteenth-century examination of ancient mounds containing Danish skulls, Foster commented that the skulls' "front teeth did not overlap as ours do, but met together like a vise."[33] Dentochronologists, notably such as Christy Turner, have developed statistical models of the rate of evolution in human teeth, which have been used to estimate relationships between population groups through time. Turner identified two major groups of dental characteristics—sinodont and sundadont—that he believed captured the evolution of dental development. Turner's model provides one method of estimating the relationship of populations through time.

Critical to such analysis is the availability of intact ancient human teeth against which the model can be calibrated. As is the case with human skulls, there have been very few human teeth discovered for the period of interest. Not only is there a scarcity of suitable remains in the New World, but there is a parallel void as well in Asian areas from which the First Americans may have originated. If Turner's dental models are correct, First American scientists could conclude that American Indians had a common ancestor in North China approximately 20,000 BP.[34] If true, this is still sufficiently broad to accommodate both the Bering and coastal migration theories. Nevertheless, it provides little insight as to whether this common ancestor of modern American Indians was the First American or rather was an ancestral source for a later arrival.

A challenge for scientists is in reconciling the findings of the different knowledge domains drawn on in the search for the First Americans. Well-matched findings based on independent observations from different scientific disciplines can considerably strengthen the plausibility of the explanatory power of a particular theoretical model. However, competition as well as cooperation across the disciplines can also introduce a level of complexity that makes the task even more difficult.

In the mid-1980s, Joseph H. Greenberg produced what was then believed to be a seminal linguistic analysis that demonstrated that the New World was populated by three separate waves of migration from Asia. About the same time, Turner produced a dentochronological study that was also considered seminal in which his findings matched the general pattern of three Asian migrations found by Greenberg. In a 1986 article, Greenberg and Turner (joined by S. L. Zegura) concluded that the dental and linguistic (and in this case also genetic) evidence "agree that the Americas were settled by three separate population movements."[35] It is a serendipitous occasion in the scientific community when observations from different intellectual domains can produce independently generated theories that are in complete agreement. It appeared that Turner's dental observations sustained Greenberg's three-part linguistic grouping.[36] Unfortunately, the Greenberg-Turner thesis was rapidly subjected, apparently with good reason, to brutal criticism. As the linguistics professor Lyle Campbell concluded, there were "many problems with these dental-linguistic correlations."[37] Most damaging was the contention that the "correlations between tooth groups and Greenberg linguistic classification . . . were not established entirely independently and then later correlated."[38] Turner had apparently arranged his tooth artifacts according to Greenberg's linguistic groupings, and then "discovered" a correlation between the linguistic and dental evidence.[39]

One of the strengths of bioanthropology in the popular imagination is that skulls and teeth can be readily understood by the public. Despite the critical First American science contributions of carbon dating in physics and DNA analysis in genetics, radioactivity and the DNA helix cannot be seen. A skull or a tooth can be held in one's hand allowing a more personal interaction with the artifact. This may, however, result in the popular perceptions surrounding the discovery of skulls and teeth as being more subject to prevailing social influences than are today's physics or genetics.

In the early 1920s, events combined to produce an amusing incident surrounding a tooth found in Nebraska. In the context of those times, which culminated with the *State of Tennessee v. John Thomas Scopes* "Monkey" trial in 1925, there was considerable division in the United States over the extent to which Darwinian evolutionary theory should be applied to humans. Against that background, a badly weathered tooth was found in Nebraska and was immediately judged to be that of a prehistoric ape-man—the "Nebraska Primate."[40] It was prominently displayed alongside other significant primate dental artifacts. Not only did this discovery support the evolutionary argument of the Darwinians, but it also gave the United States an important place alongside Europe and Africa in producing evidence of having hosted some form of human antiquity.

By 1925, the time of the Scopes trial, accounts of the Nebraska Primate were appearing in the popular press. In the February 19, 1925, *New York Times*, the tooth was reported to have been destroyed when it was handed to an assistant at the American Museum of Natural History with the comment "be mighty careful, that tooth is worth a million dollars." This so rattled the assistant that he dropped the tooth, which shattered. Despite this seemingly catastrophic destruction of a priceless artifact, it later was determined that the tooth was only that of a pig, and not a primate.[41] Nevertheless, for a time the tooth figured prominently as proof of an ancient origin for the First Americans.

One might believe that it would be difficult to find a reputable scholar that argues that there is a legitimate place for race distinctions in the search for the First Americans. Such is not the case. Ironically, among those articulating a need for a continuing role for race are American Indians—the very people who have seen bioanthropology used, unconsciously or otherwise, for racist purposes against them. In a May 2010 article titled "Working Together on Race and Racialism in American Archaeology," the noted American Indian historian Roger C. Echo-Hawk wrote that "most racial Indian archaeologists will stay deeply committed to race. In the story that Indians tell one another around their electronic online campfires, Indian racial identity serves as a unifying source of power in advancing both racial and community social agendas."[42] Echo-Hawk believed that racial identity can be used as a source of social power, and that American Indian anthropologists need have little recognition "that science has let go of race in the teachings of the academy. Race ought to continue to serve as an enduring biological truth."[43] This would appear to put bioanthropologists in the delicate position of having to operate not only in an environment where the mainstream American narrative holds that racial classifications have become pejorative distinctions, but also having to examine American Indian culture in such a way that they "must peacefully cultivate neutral ground, neither affirming race in a misguided anti-science way, nor opposing the ability of people to practice treasured belief systems."[44]

If cultures are free to make up their existential myths independent of science, and if cultures are also free to create self-identities such as race independent of science, is the culture of First American science not also free to examine these myths and identities? The difficulty for bioanthropology is that it has not yet successfully found a way to address this connection between race and the materiality of First American science that is acceptable to scientists, mainstream American culture, and the modern American Indians that are many times the subjects of their studies.

## CHAPTER 8

# The Other Sciences: Genetics, Linguistics, and Physics

A s (self) important as archaeology and anthropology are in the search for the First Americans, many of the advances of the last half century have come from the contributions of other scientific disciplines. This chapter examines these contributions as well as the inherent limitations of genetics, linguistics, and physics.

## Genetics

The search for the First Americans is, at its most fundamental level, a search for relationships. Even assuming that it was possible simply to find the physical remains of the very first human to set foot in the New World, and then to substantiate that the remains were indeed the primal First American, the most interesting scientific issues and popular public interests would still be unresolved. The critical discoveries would be those addressing human relationships, the ties of culture, the ancestral tree.

The science of archaeology is suited to the discovery and description of material artifacts. In some situations, it may be able to identify a similarity between tools used at different times and places. It is not, however, well situated as a science to make definitive determinations as to whether that similarity resulted from technology transfer between otherwise unrelated cultures, from independent development, or was nothing more than the discovery of two artifacts used by essentially the same culture at two different times and locations. Even when relevant information can be gleaned from these material artifacts, it is not the provenance of the artifacts that is ultimately of interest in the search for the First Americans, but rather what these items can suggest about their human creators and users.

What First American scientists and the popular American imagination ultimately seek are the cultural ties associated with the First Americans: where did they come from, how did they get here, and are their descendants still here? David Meltzer correctly stated that forming an estimated date for the arrival of the First Americans "cannot be bracketed on a priori grounds, archaeological or otherwise." Yet Meltzer also proclaimed that "the timing of the entry of people into the New World is a question that can be answered only by doing *archaeology* in the New World."[1] Despite Meltzer's contention, the science currently best situated to address these issues is genetics. Archaeology is much less able than genetics to calibrate the distance between human populations. Artifacts are proxies for drawing a relationship between cultures. Similarity in material artifacts is insufficient to confirm descendant ties between two populations. Moreover, artifactual dissimilarity is also not sufficient to dismiss biological ties. Genetics, while not perfectly descriptive, is able to offer a path to making judgments about direct relationships between the humans themselves, not between their cultural residue.

As discussed previously, bioanthropology can provide an estimate of human relationships based on the morphology, the structural traits, of human remains. However, given the scarcity of viable candidates for the remains of the First Americans, genetics provides scientists a mechanism for determining human relationships even when the remains are insufficient to provide morphological information. "More and more, we will see a lot of genetic information coming from fossil remains in which very little morphological information exists," according to Svante Pääbo, a Swedish biologist who is director of the Department of Genetics at the Max Planck Institute for Evolutionary Anthropology in Leipzig, Germany.[2] He was speaking with regard to the results obtained from only a finger of what appears to be an archaic Denisovan human, a third Homo sapiens in addition to Neanderthals and modern humans, called X Woman, discovered in southern Siberia. A more recent set of Denisovan bones was found in Tibet, China. The initial conclusions were based on mitochondrial DNA, and tests have been run on nuclear DNA. In addition, human hair can also be used to sequence the ancient human genome according to Eske Willerslev at the University of Copenhagen. Based on an analysis of the hair, Willerslev concluded that a body found in 1986 buried in Greenland ice for four thousand years is related to people currently living in eastern Siberia, and he is now looking at the hair from ancient mummies in the Western Hemisphere.[3]

Archaeologists view themselves as being centered in the past, and, therefore, are to be accorded the status of arbiters of antiquity; they "only see what the Pleistocene left behind."[4] To the extent that this is a valid characterization, it would appear to give archaeology a privileged ontological position vis-à-vis the other sciences in the First American debate. Only it would be able to make conclusive determinations. Yet, an epistemological domain, including archaeology, can only perceive reality within the boundaries created by the methods and value systems extant in that field. Genetics offers a capability not available in archaeology to precisely fix human biological relationships and, therefore, from these biological ties to make projections concerning cultural affiliations. "Archaeologists have trouble enough deciding which stone tool assemblages belong together; we are even harder pressed to link archaeologically detectable patterns with those identified among modern languages or genes."[5]

The modern preoccupation with DNA is a daily occurrence. It is not only central to our biological identity, it is also a part of our political debate, our health care research, our judicial system, and a perceived path to a better future. DNA, however, is increasingly also a useful tool in attempting to clarify our past. Genetic evidence has documented a relationship between modern Asians, American Indians, and early Americans. By comparing ancient DNA (aDNA), when it is serendipitously available, with modern maternal mitochondrial DNA (mtDNA) or paternal non-recombinant Y-chromosomal DNA (NRY) material from current populations, geneticists can not only draw relationships between population groups but also project such relationships into the distant past.

According to Tom Dillehay, genetics has "become a powerful tool in suggesting the origins of the first Americans."[6] However, not all experts agree on the conclusions that genetics-based data might produce. On the one hand, leading scholars at Texas A&M University and at the University of Kansas believed that "current genetic evidence implies dispersal from a single Siberian population toward the Bering Land Bridge no earlier than about 30,000 years ago (and possibly after 22,000 years ago), then migration from Beringia to the Americas sometime after 16,500 years ago. The archaeological records of Siberia and Beringia generally, but not conclusively, support these findings, as do archaeological sites in North and South America dating to as early as 15,000 years ago."[7] On the other hand, Dillehay suggests that "collectively, the genetic data suggest that all major lineages found in living Native Americans are represented in modern-day populations in northeastern Asia and that there were 1 to 4 separate migrations from somewhere in this region to the Americas."[8] As can be seen from these two examples, reputable scientists can, and do, make divergent conclusions based on the same genetic data base.

The disbelief that all human "races" with their variations could be traced to a single source was a part of the nineteenth-century debate surrounding the nascent fields of evolution and bioanthropology. Charles Lyell was concerned with reconciling the perceived differences in intellect between the "Negro and European" against theories of "a unity of origin" of the races that had subsequently "diverged from the common stock."[9] John Wells Foster concluded that "there will be found continuous and uninterrupted causes which shall explain all the diversities in the different branches of the human family, without the necessity of resorting to independent creations."[10] It was the genetics of the twentieth century that began to provide meaningful answers to these nineteenth-century questions of human origin and divergence.

What exactly is the basis of the genetics that First American scientists use in their search? There are two primary tests for using human DNA to determine an ancestral relationship. The first is mitochondrial DNA (mtDNA), which is non-nuclear and traces a single-nucleotide polymorphism (SNP) and its associated mutations that are passed down maternally. The other is based on the Y-chromosome (Y-DNA), also called non-recombining Y (NRY), and provides documentation on Y-based SNP and its mutations that are passed down paternally.[11] The rate of mutation can be estimated and, therefore, provides a sort of genetic clock that can be used to estimate the convergence toward a common ancestor. Based on the central tendencies of commonality across these mutations, geneticists have created populations called haplogroups.

Having two separate genetics clocks, however, is a problem of its own sort. Since the rate of mutation of mtDNA is faster than that of Y-DNA, it sometimes is considered to provide a more accurate calibration, much like measuring time in days instead of years. However, because there are more haplogroups identified in NRY, it provides a finer discrimination tool and is considered "the most informative haplotyping system."[12] But the differing mutation rates for mtDNA and Y-DNA can result in different estimates. As a result, it is difficult to assign one-to-one correspondences between the two haplogroup structures, which might permit accurate calibrations between the two systems.[13] Moreover, the two approaches give different answers for genetic relationships. For example, one NRY analysis in South America would lead us to believe that there is such genetic consistency that virtually the entire indigenous population can be represented within one haplogroup. However, the results of an mtDNA analysis are substantively different from that of NRY and indicates that there are four different haplogroups that have provided substantive ancestral DNA.[14] Both answers cannot be correct, and it is possible that neither is entirely accurate.

According to James Chatters, the mtDNA of modern American Indians typically falls into five distinct haplogroups. Three of those haplogroups—A, C, and D—are also found today in Siberia. A fourth—haplogroup B—is typical in modern southeast Asians. Finally, haplogroup X—typically Eurasian—is also found in modern American Indians.[15] It should be remembered that these associations are not between *First* Americans and *ancient* non-New World inhabitants, but rather between *modern* American Indians and *modern* Old World inhabitants. The difficulty comes in projecting these relationships back through time. According to Meltzer, the ultimate reward is "to firmly link the most ancient and most modern Native Americans, determine the number of migrations, from whence and where they came, or even the route(s) traveled."[16]

Because American Indian haplogroups are not distributed equally across the New World population, geneticists can draw some conclusions from the variations in the distribution.[17] "The pattern is suggestive of a sequence of migrations, a first one with D and C, a second one dominated by B, and a third consisting only of type A individuals. When haplogroup X came cannot be surmised from the geographic patterns, although there is some indication it may have first arrived on the northwest coast of North America."[18] Based on genetic analysis, Theodore Schurr concluded that "the American progenitors left their homelands between 24,000 and 35,000 years ago."[19] This is some ten to twenty thousand years earlier than either the Clovis or Monte Verde cultures that have been documented by First American scientists. The roughly one dozen ancient human remains in North America from which mtDNA has been analyzed have produced only haplogroups B, C, and D. "Haplogroup A, the most common haplogroup among living native North Americans, has not yet been found in remains older than 6000 B.C.," which supports contentions that modern American Indians are descendants from a later migration from Siberia and not the First Americans.[20] The subtlety of this analysis, both genetically and politically, is further compounded by aDNA studied by Frederika Kaestle at an ancient archaeological site in Windover, Florida. Kaestle concluded that the remains are "not A, not B, not C, not D, and not X. We've been able to prove what they're not, but we don't know what they are."[21] Conclusive answers are elusive, but new genetic evidence continues to be discovered that generally supports some version of a Beringia migration hypothesis.[22]

Although the genetic analysis supporting the search for the First Americans has largely centered on humans, there are other life forms that have coevolved with humans over the millennia and can be considered markers for a human presence. As an example, the HTLV (Human T-Lymphotropic Virus) retrovirus has been used in an attempt to identify prehistoric migrations. If the results of that study are to be believed, "Japanese sailors were the first people to reach the Americas, millennia before Siberians wandered across the Bering Strait."[23]

Regardless of whether the genetic analysis has been performed on the DNA of humans or on other life forms, genetic science has inserted itself in significant ways into the search for the First Americans. It has identified ties between modern American Indians and current groups in Asia. It has also provided evidence of multiple migrations into the New World. What it has not done is identify either the First Americans or locate their Old World progenitors. As Meltzer has commented, "Genetic studies thus far cannot confirm conclusively how many major grouping there are of modern native North Americans, much less the presumed number of migrations."[24] The difficulty in reconciling the results of genetics, bioanthropology, archaeology, and geology is that genetics suggests an earlier arrival date into the New World for the First Americans when compared with archaeological results based on material artifacts. Comparison of Asia-New World migration chronologies suggested by different theories of human migration from Siberia to the Western Hemisphere indicates a potentially earlier presence in the Americas than material archaeological findings have discovered to date.[25]

In the search for the First Americans, genetics suffers much of the same fate as bioanthropology: it is impossible to isolate the practice of genetic science from the social context of politics, race, and cultural identity. As an example, scientists at Arizona State University took blood samples from several hundred Havasupai Indians who live in the western end of the Grand Canyon. The Indians had given permission for the general study of medical disorders, and apparently many thought that the blood samples were to be used only for a study of diabetes. "When they learned years later that the DNA samples had been used to investigate things they found objectionable, they felt betrayed. Researchers had . . . traced the tribe's ancestral origins to Asia, contradicting traditional stories holding that the Havasupai had originated in the Grand Canyon."[26]

There are two principal issues involved in the Havasupai controversy. First is the potential ethical issue of performing genetic research without the informed consent of the human subjects. This concern is not confined to the search for the First Americans. The second, and much more subtle, issue is what amounts to a confrontation between genetics as a science and the substantive challenge that it can present to the cultural beliefs of American Indian tribes. The Native American Graves Protection and Repatriation Act (NAGPRA) gives American Indian tribes legal control of human remains, regardless of their age, found on their or federal lands when a cultural or geographical affiliation can be demonstrated between those remains and a particular American Indian group. In recent years, genetics has increasingly been offered as a means for determining such affiliation when other evidence is ambiguous. As demonstrated in the Havasupai example, not only can genetics provide strong evidence as to a biological affiliation, it can also directly challenge the existential myths of an American Indian culture.

In 2000, the secretary of the interior—who was given statutory authority by NAGPRA to make cultural affiliation determinations—ordered that DNA tests be conducted on some human remains. This proved to be a controversial directive for both the American Indians and the First American scientists involved. The American Indians objected because DNA testing was an affront to their religion and a belief that cultural identity was based on more than simple genetics. Interestingly, some of the scientists also objected, arguing that culture is learned and therefore has no relationship to biology.[27] As Vine Deloria Jr. has pointed out, one of the most controversial problems today between First American scientists and American Indian groups is the application of genetics as a proof of tribal affiliation.[28]

## Linguistics

First American linguists are largely trapped in the evidentiary present. In the absence of any evidence of written languages for both the First Americans and even pre-Columbian modern American Indians, linguistic theories can only be validated through comparing known languages of the past few hundred years. If linguistic tools are of value when applied against the written and spoken records in Europe and Asia, they are of considerably less value in the context of the human history of the Americas where there are few written records, and none associated with the First Americans. Moreover, in attempting to generate a linguistic tree for American Indians, some scholars contend that modern American Indians are not direct descendants of Paleoamericans; consequently, a linguistic tree that cannot examine a missing branch is of little value. Finally, there is no independent reality against which linguists can recalibrate either their hypotheses of Paleoamerican linguistic relationships or the clock on the rate of linguistic evolution. "One cannot recover fossil languages, at least not until the development of writing."[29] Accurately estimating a linguistic tie between two cultures that is chronologically inaccurate by, say, two thousand years can be the difference between a simply interesting result that ties together two neighboring cultures in Siberia, as opposed to a theory-invalidating result that ties together a culture in Siberia with one in North America. Nevertheless, there has been an active linguistics effort that searches for clues in the languages of modern Native Americans that could lead to an ancestral linguistic convergence that might be associated with the First Americans.

Early in American history there was a sense that linguistic evidence might have a bearing on the source of the First Americans. As a part of Thomas

Jefferson's 1784 excavation of a mound that was determined to have been the product of American Indians, he concluded that linguistics dictated that American Indians and the people of northern Asia had a common origin, and that American Indian languages must be more ancient due to their greater diversity.[30] However, Jefferson's view was not universally shared. Benjamin Smith Barton (1766–1815)—a professor of medicine, natural history, and botany at the University of Pennsylvania—believed that the diversity of American Indian languages was such that they could only have separated from a common origin at a time that predated Bishop Ussher's Earth-creation date of 4004 BCE.[31]

By the nineteenth century, First American scientists were even more confident that linguistics could provide a powerful tool for tracing relationships between modern and ancient cultures. In 1874, John Wells Foster commented that "The study of languages affords a reliable guide in tracing the migrations of tribes, even where they have become intermingled with other tribes. In the social relations thus established, there would not result a total obliteration of the language of the one tribe, but certain words and forms of speech would be adopted and perpetuated."[32] When applied in the context of European languages, Foster believed that they "can be traced back to the Sanskrit as an approximate common source."[33] Ignatius Donnelly echoed Foster's confidence in tracing modern Europe's linguistic roots. "Philology is yet in its infancy, and the time is not far distant when the identity of the languages of all the Noachic races will be as clearly established and as universally acknowledged as is now the identity of the language of the Aryan family of nations."[34] Yet the nineteenth-century ground in which linguistics proved fertile for documenting European linguistic history was apparently barren in North America for the search for the First Americans. Foster concluded that "the language of the American Indian throws no light upon his origin," and that linguistic efforts to trace the origin of American Indians were "utterly futile."[35] He could find no elements of the ancient Hebrew, Welsh, Hindu, Phoenician, Chinese, or Scandinavian languages that should be related to modern American Indian languages if those civilizations had supplied the ancestors of modern American Indians.[36]

As the fields of anthropology and linguistics each developed into professional academic disciplines in the twentieth century, Edward Sapir (1884–1939) was among the first to combine the domains in his search for the First Americans. Because of the great diversity in American Indian languages, Sapir believed the explanation could be either that American Indians had been in North America for such an extremely long time that their languages had had sufficient time to diverge, or more likely that the divergence resulted from a "series of movements of linguistically unrelated peoples."[37]

Understanding the application of linguistics in the search for the First American is important for what it reveals in how the science is practiced. It has frequently exposed the boundary lines between the scientific domains involved in the search. In 1987, Joseph H. Greenberg (1915–2001) published what was a then-seminal work titled *Language in the Americas*.[38] In it, Greenberg concluded that the myriad languages of the indigenous North American people could be resolved into only three linguistic families, which he named Amerind, Na-Dene, and Eskimo-Aleut.[39] It was generally accepted that these linguistic groupings were the result of three separate migrations from Asia, most likely via Beringia.

Greenberg's thesis was presented in an increasingly interdisciplinary First American scientific environment that saw in the 1980s a series of papers on dentochronology by Christy Turner analyzing the relationships of American Indian and Asian teeth that supported Greenberg's three pulse migration thesis.[40] For Turner, the issue has now become that his findings may have also been unduly influenced by Greenberg's three-migratory-wave thesis.[41] Moreover, Greenberg's results have now been challenged as too simplistic. David Meltzer commented that "it is probably no more realistic to infer Pleistocene migration routes to North America by the number and distribution of modern language groups than it would be to infer Hernando de Soto's route by looking at the number and distribution of Spanish dialects in the Southeast today—and at least we know that de Soto spoke Spanish."[42]

Linguists are confronted with the possibility that the results of their work may ultimately be irrelevant in the search for the First Americans. As was indicated in a Smithsonian Institution report, "No American Indian language is derived from an historically known Old World language."[43]

## Physics

If archaeology or anthropology were to disappear, the discipline of physics would scarcely notice. If physics, however, were to disappear, First American science would largely revert to relying on the analytical tools of the nineteenth century for dating artifacts. In the search for the First Americans, one method of obtaining observational data is to document the environment of an artifact of interest; that is, for example, to situate the material within its known geological context. The best data, however, is obtained directly from the artifact. For the past sixty years, the principal means of dating human and other life form remains has been what is popularly called carbon dating.

In 1947, the chemist Willard Libby established that the steadily predictable decomposition of the carbon 14 isotope into nitrogen 14 could be used to date ancient human remains. In 1960 Libby was awarded a Nobel Prize for his efforts.[44] Carbon 12, 13, and 14 are taken in by plants through photosynthesis and then ingested by animals in relative proportion to that found in the environment and in their specific diet. Although carbon 12 and 13 are stable, carbon 14 decays. Upon death, carbon 14 is obviously neither absorbed during photosynthesis by plants nor ingested by animals. However, the carbon 14 that was already in the then-dead plant or animal continues to decay into nitrogen 14 at a rate such that after every 5,730 years, 50 percent of the remaining carbon 14 is gone. Based upon this known rate of decay, an estimate can be made as to length of time that has transpired since death. Archaeologists typically provide these dates in carbon dated years called "BP," that is **b**efore **p**resent—with the present being standardized as 1950, when carbon dating began.

There are, however, limitations to using radiocarbon dating. The most advanced technique currently used by physicists is called accelerator mass spectrometry (AMS), which counts not only the carbon atoms that are decaying, but also counts all of the carbon 14 atoms present in order to make a very precise estimate of the decay ratio. Application of a technique as sophisticated as AMS is constrained in that it is expensive, so it cannot be used routinely. In addition, any "sample that is 35,000 years old will have only 2 percent of its original carbon-14 left. Even a tiny bit of a recent contaminant is enough to skew the radiocarbon date seriously."[45]

An additional difficulty in using carbon dating is that the amount of carbon 14 first in the atmosphere, then in plants, and subsequently in animals has varied over the millennia. By analyzing tree rings and the carbon trapped in ice caps, scientists have been able to provide a calibration for the radiocarbon dates that permits translation into an approximation of a chronological year. These dates can change by thousands of years, with the commonly accepted Clovis-first date being approximately 11,500 radiocarbon year BP (i.e., 1950) or thirteen thousand chronological years ago.

The difficulty is further compounded in that the ingestion of carbon is a function of a person's diet. People who have a diet high in marine food sources have a lower rate of carbon ingestion than do those whose diet is primarily of terrestrial origin.[46] In addition, over the past ten thousand years, human activity has pumped vast amounts of dead plant carbon into the atmosphere through the burning of wood and hydrocarbons.[47]

The question then becomes: when is a year not a year? It is not a year when attempting to calibrate chronological years with those of the different disciplines used in the search for the First Americans. This complexity of the dating problem was aptly captured in an article by Ted Goebel, the first footnote of which was as follows: "All ages are presented as ka (thousands of calendar years ago). Dates relating to genetic events are in calendar years based on coalescent methods. Dates relating to archaeological events are derived by calibrating radiocarbon ages. Radiocarbon dates younger than 21,000 $^{14}$C years ago were calibrated with Calib 5.0.1 (IntCal04 curve); older dates were calibrated by using CalPal Online (CalPal 2007 HULU curve)."[48]

What Goebel's methodological caveat means is that a chronological year does not equal a genetic year, which does not equal a radiocarbon year. Furthermore, even within a single discipline such as genetics, a mitochondrial DNA (mtDNA) year does not equal a non-recombinant Y-chromosomal DNA (NRY) year. As Goebel continued, "The differences in calculations are the result of several issues, including potential variation in mutation rates, variable and sometimes circular techniques of calibrating coalescent times to calendar years, time-dependency of mutation and/or substitution rates, and effects of genetic drift on the original founding population."[49]

The chronological year has the merit of being easily understood and universally observed, and having been constant within a few seconds since well before human existence. It is also the standard against which these various "scientific" years attempt to calibrate themselves. Yet the disparities can be quite large. Geneticists have their problems with establishing chronologies. Geologists have their own set of issues. Carbon dating is a function of the mix of dietary intake and atmospheric gases, which have varied substantially over the period that covers the peopling of the Americas. This all produces a temporal incongruity where fifteen thousand radiocarbon years ago (i.e., before 1950) equates to over eighteen thousand chronological years ago (i.e., before today). At least that is the discrepancy that scientists currently believe is correct based on analysis of tree rings and ice core samples.

As advances occur in the ability of physics to provide other mechanisms for dating nature, more complications appear to arise. For example, by determining a point at which radioactive thorium 230 and radioactive uranium 234 reach an equilibrium point in their decay, it is possible to use a different physical clock to calibrate the carbon dating technique. Unfortunately, uranium-thorium dates raise questions as to the believed calibration between radiocarbon and chronological dates. First American scientists were confronted with one of their members arguing that they faced the prospect that "the actual time of Clovis expansion may have been about 11,500 B.C., not 9,500 as indicated by radiocarbon."[50] The point is not whether that proposal prevailed, but rather that what is believed known may be proved wrong. Such is the very nature of science in general.

Other dating-associated problems arise. As discussed above, it is known that radiocarbon years can vary depending on annual variations in the natural creation of the carbon 14 isotope that is taken in by plants and animals. Consequently, it was necessary to find a reliable mechanism to calibrate radiocarbon years with chronological years. A method of choice has been provided by the science of dendrochronology—the counting of annual tree rings, a practice known at least as far back as Leonardo da Vinci—which has provided a calibration back to the start of the Holocene approximately ten thousand years ago.[51] One of the first reported instances of the use of tree rings in support of the search for the First Americans was in 1788 by Manasseh Cutler to date American Indian mounds in Ohio.[52]

As valuable as tree ring counting may be for some purposes, its use as a dating mechanism is generally conceded to extend only through the past approximately ten thousand years. Regardless of who they were or when they came, it is clear to First American scientists that the migrations to the New World happened before that time. Consequently, the calibrations between radiocarbon and chronological years become increasingly more suspect the farther back scientists venture into the critical period of interest in the search for the First Americans.

Another method for calibrating radiocarbon dating results with chorological years is the counting of layers of ice cores, typically in Greenland. However, it too has its share of controversies. One debate is centered on the extent to which the Earth's climate is relatively stable, the uniformitarian hypothesis, and that annual layers of ice are relatively thin. Others counter that the climate is not stable, and the layers can be quite thick. The irony of using ice cores to calibrate radiocarbon dating is that ice cores themselves have to be calibrated and scientists use such measures as deep sea cores and volcanic acidity spikes in the ice cores.

Physics has contributed other techniques. Some are useful to the period of interest in the First American search. One is thermoluminescence dating, developed in the 1960s, which measures the amount of light released in crystalline materials when heated in a laboratory. The amount of light released is proportional to the time elapsed since the material was formed. Thermoluminescence is useful for dating these minerals between several thousand and a few hundred thousand years ago. It has been used effectively to document the forty-thousand-year-old presence of humans in Australia.[53] The decay of potassium 40 in minerals into argon 40 can only provide estimates beginning several hundred thousand years ago, since it takes over a billion years for even half of a sample of potassium 40 to decay into argon 40.

Despite the several weaknesses cited above to using radiocarbon dating in the search for the First Americans, a cottage industry has arisen offering dating to those who can afford the tariff. Sponsored by the Society for American Archaeology, commercial vendors set up booths offering services at the annual SAA academic convention with price lists for radiocarbon dating based on either AMS or liquid scintillation counting, and, of course, to be performed under the supervision of a staff of PhDs.

The tools of physics have become critical in the evidentiary debates surrounding the search for the First Americans. Yet, as Edwin N. Wilmsen originally wrote as early 1965, "radiocarbon dating is no miraculous tool; sound stratigraphy, observation, and correlation with geological and paleontological evidence become even more necessary as more precision in dating becomes possible."[54]

# CHAPTER 9

# Laboratories and Museums

I N 1784, THOMAS JEFFERSON dug into a mound located on his land and concluded that it had been built by the ancestors of American Indians. When it had been built "was a matter of doubt."[1] Historians frequently refer to Jefferson's effort as being the first instance of a specific American excavation for archaeological purposes. "Jefferson's pioneering methodology—the first excavation designed not to recover artifacts but to solve an archaeological question—was sufficiently ahead of his time that it had virtually no real impact on subsequent work for at least another century."[2] Thus began the creation of the First American scientific laboratory and museum in the search for the First Americans.

## Laboratories

For the modern First American scientist, much of the world's land mass, and some parts of the oceans as well, constitute the potential laboratory. Many scientists have a fixed point, a known location, to which they can repeatedly return to test their hypotheses. Archeologists must first create an expectation, a theory, as to where to even situate a laboratory. Where are they most likely to find material artifacts, to make meaningful observations? In a cave? In a desert? Beneath a body of water? Where there are, or rather were, flora and fauna of particular types? As an example, the accidental initial discovery by an amateur of the Clovis cultural tool kit occurred in the high desert area of the American Southwest and so the location for an important First American laboratory was serendipitously created in a dusty arid field that, for a while, was shared with a gravel mining operation.

Since Jefferson's time there have been many more sites, including important ones in South America. The more famous tend, naturally, to be those that have produced novel results as well as those that were more contentious. As James Adovasio pointed out, "the cavalcade of loser localities was taken by many archaeologists as proof that no one was or could have been in the New World before the fluted-point makers [Clovis]. Such a belief is, of course, not science. And it is not logical. It is, in fact, more like a religious dogma."[3]

The goal of the First American laboratory is to produce and examine material artifacts from which culture-defining statements can be made. The archaeologist could, ideally and unrealistically, hope to find a living, breathing potential First American. That being an impossibility, the hope is to find a first order derivative: the remains of a First American. However, typically, what is found is a second order derivative: the material artifacts left behind by no-longer existing First Americans. Further compounding these difficulties is the reality that only a small subset of the cultural tool kit is likely to survive. Flaked stone tools are frequently the only material evidence recovered from First American archaeological sites.[4] As a result, archaeological theory associated with the First Americans is irretrievably linked to methodologies that are structured toward finding stone tools. "Their [nondurable artifacts of wood, bone, or fiber] absence has lent undue importance to the relatively common stone tools. This bias in what has been preserved and what hasn't has in turn helped create (and sustain) an image of Late Pliocene and early Pleistocene technology that is not only wrong for that time but for virtually all later periods. . . . It is virtually certain that stone was always a minority element in their technological suite."[5] Moreover, it is believed that stone tools are historically the result of a masculine endeavor, overemphasizing the hunting/killing aspects of First American society; the perishability of materials such as wood, basketry, and cloth results in an unrepresentative view of that society and the roles of women and children.

First American scientists face a particularly daunting evidence-by-proxy challenge. Their task is to develop plausible theories regarding prehistoric cultures that might well have no modern progeny. In the absence of a cultural road map, scientists typically prize burial sites for the information that a cadaver and its associated funereal artifacts provide. For the First Americans, there are few such remains, and those that exist are typically subject to contentious arguments over physical control of the remains themselves. That then leads scientists on a search for artifacts of any type. Often, the only humanly modified material that has survived through the millennia is stone. However, as valuable as stone may be as an enduring talisman of a cultural past, it may well not be the predominant material tool employed. First American scientists, such as Dennis Stanford and Richard Morlan, believed that pre-Clovis cultures might have predominantly used bone rather than stone in their cultural tool kit. Such a contention is purportedly supported by thousands of bones found in the mudflats of the Old Crow River in the Canadian Yukon.[6] Finds such as Old Crow are rare. The

Stanford-Morlan thesis may well be correct, and their logic of stone being pref-
erentially preserved by nature has a certain intuitive plausibility, but current
evidence is not sufficient either to prove or disprove it.

The determination as to how to construct any particular laboratory is critical
for the practice of science in the search for the First Americans. For an archaeologist as well as for a bioanthropologist, the scientific act of discovery is first an act
of physical discovery, with the initial problem being the selection of a location on
the Earth. To obtain funding from a sponsor for an archaeological excavation,
it is necessary to demonstrate a reasonable probability of success—the measurement of which is typically against generally prevailing theories. When there is
a theory, such as Clovis-first, that is believed to be well-confirmed, a proposal
to search (i.e., to situate a laboratory and make observations) for a pre-Clovis
site could be perceived as baseless with the resulting consequence that experiments were typically structured and funded to confirm repeatedly the reigning
Clovis-first theory and rarely around a meaningful effort for its refutation. With
the laboratory existence itself being a direct reflection of a theoretical model, the
observations become increasingly theory—and even observer—dependent. As
an example, Monte Verde in Chile and Meadowcroft Rockshelter in Pennsylvania—two initial challenges to the long-held Clovis-first paradigm—were neither
discovered by a scientist nor the result of a specific search based on a theoretical
model. Rather, they were each the result of a serendipitous sighting by a nonsci-
entist intent on other matters.

What constitutes a laboratory, the context of scientific observation and experimentation, is less clear in the search for the First Americans than might be
evident in an investigation into particle physics or the development of pharma-
ceuticals. "Since the seventeenth century, the laboratory has come to be recognized as the preeminent site for making knowledge in the experimental sciences.
It straddles the realms of private seclusion and public display, and calls for means
of managing the transitions between them."[7] While philosophers of science may
engage in debates concerning the relative merits of induction and deduction as
methodologies for drawing conclusions about the nature of the physical world,
these same issues play out in a much less esoteric fashion in the daily lives of First
American scientists. Is the preferred experimental approach to uncover evidence
and then create a theory with some explanatory power, or to establish a working
theory and then test it by seeking observations, typically in a laboratory, that
confirm or falsify the theory? In the practice of science in the search for the First
Americans, holes are not dug at random. As Ian Hacking observed, a scientist
must always begin with some expectation. "We should not . . . underestimate the

pre-theoretical role of invention and fiddling around."[8] First American scientists, in particular, must have some idea of how the material world might produce evidence and what such evidence might look like.

For the First American scientist, there is not only the necessity of identifying a physical location for the laboratory, but there also must be a reasonably accurate temporal assessment. Depth into the Earth translates into elapsed geological time. Digging deeply is resource intensive, but failure to explore to an adequate depth precludes discovery of artifacts relating to particular periods in human history and, therefore, excludes examination of entire classes of potential theories. The location of an archaeological laboratory can be situated in such a fashion that the earliest artifacts that might be discovered are those of the Clovis era; without an accepted theoretical model that anticipates a *pre*-Clovis culture, in the absence of a purely serendipitous accident, the material artifacts that form the basis for archeological discovery will not be found because the excavation will be of insufficient depth or in the wrong location.

In addition to the difficulty in determining precisely what constitutes a laboratory for the First American scientist, there can be little doubt that, wherever such a laboratory *could* be, the possibilities are disappearing. Just as we modern humans build our social networks in areas associated with sources of food and water, facile modes of transportation, and favorable environmental conditions, so did our ancestors. Access to productive archaeological laboratories would, in many cases, entail destruction, or at least significant disruption, of elements of modern society. The opportunity costs are sufficiently great that vast populated areas of the Earth's land mass are off limits to archaeological exploration. Conversely, that same human activity—the creation of roads, cities, and the other trappings of civilization—that creates such a physical barrier for modern archaeologists, has also already destroyed archaeological artifacts.

The theory dependence of observation is central to the First American debate. If scientists situated their laboratories in such a manner as to give discovery of Clovis artifacts the only realistic outcome of the experiment (i.e., the archaeological excavation), then the opportunities are greatly decreased for archaeological results that disconfirm a Clovis-first theory. Although methodological rigor and advanced technological equipment are important in all scientific domains, in archaeology, the theory-laden nature of the location of the laboratory is of especially critical importance.

With the typical within-domain difficulties described above, attempts to situate a laboratory frequently encounter irreconcilable differences between historical discovery and heritage preservation. Situating a laboratory and conducting an experiment that searches for First American evidence confronts the reality that many of the most logical locations are on properties owned either by American Indian tribes or by the US government, which, in either case, triggers federal laws and regulations that accord equal status to cultural preservation over the interests of science. Despite the judgmental positions taken by the advocates on either side—see the discussion elsewhere on the Kennewick Man—both science and heritage preservation have, upon occasion, been culturally hegemonic tools tainted by moral ambiguities. Examples range from Stalin's prohibition against the practice of "Jewish" science to US prohibitions on stem cell research. There may, however, be instances when a normative constraint on the practice of science may provide a balance, not a proscription, for the conflicting cultural authorities of heritage and science. In *The Leviathan and the Air-Pump*, Steven Shapin and Simon Schaffer drew a direct relationship between social order and the methodologies considered acceptable in the practice of science. There is no question that the imposition of social constraints on what constitutes an acceptable laboratory impacts the practice of First American science while giving some protection to the myths associated with American Indian heritage preservation. What constitutes an observation, a fact, is an outcome produced by a social negotiation. The difficulty in completing a successful negotiation as to what constitutes an observation is that myth and science—each of which has established a form of cultural legitimacy in the search for the First Americans—offer mutually conflicting standards as to the location of a laboratory. The gold standard of technological rationality becomes the mass spectroscopy of carbon decay, the genetic analysis of migrant haplogroups, and geospatial imagery. Instruments routinely employed in physics increasingly bestow their credibility on First American science. New types of instrumentation, however, can also provide an opportunity for a theoretical flexibility, which can challenge the social validation of outmoded models.

The so-called First American laboratory locations are plentiful. A short and nonexhaustive list of potentially significant First American archaeological sites—legitimized, refuted, or unresolved—must include Clovis, New Mexico; Meadowcroft Rockshelter, Pennsylvania; Cactus Hill, Virginia; Monte Verde, Chile; Page-Ladson, Florida; Cerutti Mastodon, California; Buttermilk Creek, Texas; Calico Hills, California; and Chiquihuite Cave, Mexico.

In his study of Louis Pasteur's nineteenth-century research on anthrax, Bruno Latour concluded that laboratories, both culturally and scientifically, are a mechanism for the generation of social power.[9] In the search for the First Americans, laboratories are a *reflection* of social power as well. Laboratory location critically constrains the potential results that may be obtained. Where excavation is conducted is mediated by several critical factors, all of which are determined in substantial part by politics. First, much of the funding for excavations now comes through the Cultural Resource Management (CRM) agendas of American Indian organizations. Second, the Archaeological Resources Protection Act (ARPA) and the Native American Graves Protection and Repatriation Act (NAGPRA) constrain where excavations may *not* occur. Finally, the location of an archaeological excavation has been decisively dependent on a dominant First American theory which, for a seventy-year period, was the Clovis-first model.

## Museums

An archaeological site serves not only as a laboratory but is in substantial ways also a museum. It typically becomes a static display of the knowledge production process rather than of the knowledge itself. The artifacts are removed, but the process is left preserved and available for inspection. It is a museum displaying a petrified laboratory methodology. In many ways, the preserved excavation sites that produced the artifacts and the museums to which they were removed are as much as a part of the First American scientific record as is any written report or artifact display.

For the Meadowcroft, Pennsylvania, and the Clovis, New Mexico, sites, once the excavation was completed, the evidence of the process itself remained as a monument to a particular methodology for years, and in the case of controversial locations, at times even for decades. As is obvious at both locations, there was a considerable resource commitment, especially at the Meadowcroft Rockshelter, to enclose the excavation site and facilitate access not only for interested professionals, but also for the public at large. Each of these preserved excavation sites, these static displays, provides the visitor the sense of being in a museum.

At the Meadowcroft Rockshelter, archaeologists have documented a clear preservation of the stratigraphic record associated with the recovered artifacts. In addition, there are hundreds of tags that permit later association of a particular artifact with the specific location documenting where it was discovered. There is no longer any active excavation underway at the site, and it is now effectively a museum. There are observational platforms, stairways, railings to guide the public, and even electronic screens showing videos of the excavation as it took place. The public is permitted access to the site only upon paying admission, and under the guidance of a knowledgeable, but nonprofessional, tour guide. Since the original work began in the 1970s, the excavation has been under the supervision of the archaeologist James Adovasio with funding for the location currently being provided by the admission charges as well as by the Senator John Heinz History Center. Unlike the Blackwater Draw location, which is associated with the well-documented Clovis culture, the Meadowcroft Rockshelter is associated with a pre-Clovis culture for which a formal scientific report has not yet been published. Consequently, the exact preservation of the archaeological site may well be critical to a later examination by independent First American scientists to confirm the artifactual integrity on which Adovasio will base his conclusions.[10]

The Blackwater Draw display for the Clovis excavation is under the control of Eastern New Mexico University, and apparently has been less well funded than that of the Meadowcroft Rockshelter. While Blackwater Draw also has viewing platforms, the building is less attractive; however, it would appear to provide functional sheltering from the environment. Exact preservation of the site is no longer an issue since the material existence of the Clovis culture is well documented not only at this location, but also at many others throughout North America. There is also a separate museum for Blackwater Draw artifacts operated by Eastern New Mexico University located nearby in Portales, New Mexico.

There is an increasing recognition of the importance of museums as knowledge production sites. As Jan Golinski has indicated, in field sciences such as the search for the First Americans, "an analysis of science as a localized construction makes no sense, since their knowledge-producing practices are not bound to any delimited space."[11] The work of First American scientists is not substantially different from Louis Pasteur's nineteenth-century incorporation of field-situated agricultural livestock as a part of his biological laboratory.[12]

With the evidentiary abundance of early human presence in Europe, the initial inclination of First American scientists was to assume a New World parallel, to assume that the human origins theories created in the European context were also appropriate for the New World. "While American archaeology was developing a more scientific approach in the years between 1865 and the turn of the twentieth century, it nonetheless lagged far behind the field as practiced by Europeans."[13] Convinced that there was a human origin analogue between the New and Old Worlds, the conceptual space of nineteenth-century American archaeological theory included room for little more than what might be called proof by European analogy—and that analogy requiring both digging and static displays.

Joseph Henry (1797–1878), the first secretary of the Smithsonian Institution, alerted travelers as to what to look for in "Indian country" with regard to remains and artifacts that might document the antiquity of human origins in the United States.[14] Even though Henry was of the nineteenth-century generation of American scientists that actively encouraged the looting of Indian burial sites in the search for the First Americans, the Smithsonian Institution oddly presents this desecration as beginning a natural continuum that resulted in the creation of the National Museum of the American Indian. According to the Smithsonian Institution, "the creation of the National Museum of the American Indian is the most recent example of the Smithsonian Institution's commitment to the increase and diffusion of knowledge about Native Americans. It is a commitment which goes back to the first secretary, Joseph Henry."[15] That Henry was committed to the study of American Indians was obvious; however, there is more than a subtle difference between looting and museum creation as mechanisms for displaying a commitment to knowledge production.

The public interest in human origins is evident in viewing the crowds that have visited the David H. Koch Hall of Human Origins in the Smithsonian Institution National Museum of Natural History.[16] Another measure of public interest is philanthropic donations, as is apparent from the $15 million contribution to the Smithsonian by David H. Koch to support the project.[17] On the same day that I saw a packed crowd at the Hall of Human Origins of the Smithsonian National Museum of Natural History (NMNH), the nearby Smithsonian National Museum of the American Indian (NMAI) was practically empty.[18] In 2018, the Smithsonian Institution reported that its facilities had a total of 28.5 million visitors. Of that number, only 1.1 million visited the National Museum of the American Indian despite its large building situated prominently on the Washington Mall between the US Capitol and the other major Smithsonian museums. In 2018, attendance at the Museum of Natural History, including the Human Origins exhibit, was 4.1 million. Acknowledging that the NMNH has many exhibits unrelated to human origins, the public was largely walking past the NMAI to visit the other attractions surrounding the Washington Mall.

Although a comparison of attendance numbers is relatively straightforward, it is much more difficult to assess the motivations behind attendance or non-attendance at any particular museum. One possibility is the content itself. In an NMAI exhibit that began in 2006 called "Listening to Our Ancestors: The Art of Native Life along the North Pacific Coast," it was stated that the "curators from each of these North Pacific Coast communities determined the content and selected the objects from the museum's collection to be included in the exhibition. They provided important information on the unique cultural context of each object."[19] It can certainly be argued that American Indians have a unique insight as to their own interpretation of their culture and its artifacts. This exhibit was apparently presented unfettered by any challenge from First American science. In a discussion with a staff member of the NMAI Resource

Center, it was confirmed that the exhibits of the NMAI are not intended to be science, but rather are to be representational of American Indian culture.[20] Based on these attendance figures alone, the American (and foreign visiting) public would appear to attend scientifically centered exhibits in much greater numbers than those representing an American Indian cultural perspective.

Museums can also play subtle, but powerful, roles in reinforcing particular cultural narratives. For example, an exhibit entitled "The Americas" at the NMAI stated that the pre-Columbian cultures of the New World have "memories from the beginning of everything."[21] This is presented as if there were an agreed-upon consensus that this was the case, as opposed to presenting it more accurately as being a part of American Indian traditions that are contested by many First American scientists.

A second example of the subtle implications as to how museums can shape the cultural narrative is contained in an NMNH video with a voiceover narration providing a chronological countdown from the beginnings of human origin to modern times. At one point the viewer is shown that by seventeen thousand years ago, Homo sapiens had occupied the entire world to include the Western Hemisphere.[22] Without clearly articulating it, in this instance the Smithsonian Institution is subtly telling the museum visitor that modern humans had already occupied the New World for four thousand years before either the Clovis or Monte Verde culture arose. While there is material evidence that Clovis was not the first culture in the New World, the timing of the arrival of the first Homo sapiens remains far from being a settled issue. Yet the Museum of Natural History provided what is by archaeological standards a fairly precise date of "about 17,000 years ago."

For First American science, the traditional view of a museum as nothing more than a static display of physical artifacts is obviously misguided. The museum-like nature of these sites where archaeological excavations have taken place is an important element in legitimizing the specific practice of science as conducted at that particular location. Conversely, First American science, as presented in traditional museums such as the Smithsonian Institution, play an important role in enlisting public sentiment in support of a particular theoretical view that results in a determination of the winners and losers in the theories associated with the search for the First Americans.

PART III

# Community

# CHAPTER 10

# Identity and Heritage Preservation

I N THE SEARCH FOR the First Americans, the sacred scientific grail is not the physical remains of the very first Homo sapiens to set foot in the Americas. As interesting as that archaeological discovery might be, what is of greater interest are the cultural findings. How did the First Americans view the world? Who are their modern descendants? Who were their ancient ancestors? What was the social structure? How advanced were their tool-making capabilities? As the archaeologist James Adovasio has pointed out, First American scientists as well as nonspecialists "want the Americas to have been inhabited anciently as opposed to only recently," which is ironic considering that "Americans pride themselves as a nation on having accomplished a great deal in a very short period of time—going from a rag-tag collection of rebels to the world's only superpower in little more than two hundred years."[1] An important element driving America's preoccupation with its first inhabitants is the constant struggle within a society to control the cultural narrative. The search for the First Americans "produces symbols, knowledge, and heritage, which give rise to awareness and consciousness of group identity and are invoked to inspire and justify collective agency. Groups wage powerful struggles over what is remembered and what is forgotten about the past."[2] Those who have the power to define the past are well positioned to influence the future.

The inability of either First American scientists or modern American Indians to clearly define the physical and cultural identity of the First Americans has continuing political repercussions today. What does it mean if a culture changes, but the people remain biologically indistinguishable, or vice versa? The ancient remains of the Kennewick Man initially provided a basis for some to conclude that the First American was physically different from modern American Indians and, therefore, there can be no cultural continuity. In turn, American Indians can and do counter this by saying that there is a cultural continuity for indigenous Americans that renders moot any biological variations through time. For them, all ancient human remains have sufficient anatomical similarity to modern American Indians to impute a common cultural heritage. Biology and culture are inherently incommensurable concepts, the metrics for which are qualitatively different, and it is difficult to make meaningful comparisons. Nevertheless, efforts to protect or reject a specific identity or heritage have resulted in both scientists and American Indians using observations drawn from one domain to justify convenient theories in the other.

The enduring challenge for First American scientists as well as for American Indians is: who owns the past? As the anthropologist Larry L. Zimmerman characterized it, "archaeologists cling to the idea . . . that the past is a public heritage. . . . Indigenous people usually disagree. Their pasts are their own, not the rest of humanity's."[3] Where do the global and the local merge? Do modern Tanzanians have some right to demand that two-million-year-old proto-human remains discovered in Olduvai Gorge be immediately reburied in secret without any study by scientists? This is not an issue in Africa; yet in the United States, existing law now provides such a power to American Indians.

Conflicts in the search for the First Americans ultimately concern power and the ability to control interpretations of the past—ownership of one's heritage—which provides political leverage for influencing modern society. According to Tamara L. Bray, "an implicit consequence of repatriation legislation is the negation of Americanist archaeology's claims to exclusive control over the interpretation of the past. The alternative epistemologies and systems of knowledge authorized by the [Native American Graves Protection and Repatriation Act] legislation challenge the privileged position of traditional academic knowledge."[4] History supports Bray's implication that First American science has been employed to support a nonuniversal set of cultural values—a particular cultural heritage. An obvious and well-documented example is the racism of nineteenth-century American anthropology. This difficulty of early First American scientists in disinterestedly approaching the study of human origins was evident in the romanticizing in the 1874 writings of John Wells Foster when comparing human capabilities to other animals. "Man is endowed with reason, by the exercise of which he makes up for all these deficiencies, repels or subdues all other animals, and places himself at the head of creation. He soars in the air above the eagle's flight; he sounds the depths of the sea beyond the fishes' range; he outstrips the pigeon in his flight over the land; his vision, aided by art, pierces the realm of space infinitely deeper than that of the condor's; and by the pressure of his fore-finger, exerting less strength than would be required to lift a pound weight, he sketches the lordly lion in the dust."[5] With such a view, it is appropriate to ask if it is possible to perform value-neutral research within the bounds of what might be considered objective science. In retrospect, Foster's cultural heritage was clearly embedded in his work, which reinforced the then-scientific wisdom of a hierarchy of races.

Even when executed in good faith and according to contemporaneous principles, science nevertheless is capable of producing fantastical myths—such as Atlantis as the source of the First Americans. Once launched into and accepted by a culture, an attractive myth dies hard. It is difficult to conclude that myths, such as those of the modern American Indian, generated through extra-scientific traditions and oral histories, are not subject to a revisionist process such as that generated within science. Control over what constitutes the past is not a new issue nor unique to American society. In *Requiem for a Nun,* William Faulkner wrote that "The past is never dead. It's not even past."[6]

Nearly two hundred American Indian federally recognized tribes have something equivalent to the Zuni's Heritage and Historic Preservation Office. The existence of such an office is at once an interesting statement not only that both heritage and history are entities that can (and ought to) be preserved, but also that the simultaneous preservation of both is a sufficiently compatible exercise that it can be accomplished within the same organization. Preservation would also necessarily imply that there is a fixed and knowable entity—whether physical or conceptual—that can be conserved. In contrast, the prevalent academic view is that any particular history is, at least partially and perhaps even substantially, a function of the values of the culture of the historians writing the history. Consequently, any evolution of a particular history depends on changes in the dominant culture as well as the discovery of new information, new facts, that can successfully challenge the dominant history.

It is difficult to envision that heritage preservation can occur in any context other than one in which counterfactuals and new interpretations are not encouraged. Inasmuch as the Zuni Preservation Office, as do counterpart organizations in other American Indian tribes, employs professionally trained archaeologists and anthropologists who presumably view and conduct their crafts as a science, this would suggest that some significant number of First American scientists are comfortable in practicing their discipline in an environment that may not subscribe to the often-debated role of science as a value-neutral interpreter of the physical world; rather, some First American scientists appear now to be overtly dedicated to preservation of one particular cultural interpretation. Heritage, which is a mixture of tradition, identity, belief systems, and physical representations, is an endeavor that they apparently believe is susceptible to preservation, and in this case, one in which scientists participate. Such socially engineered preservation offices are institutions specifically chartered to promote a particular set of cultural values.

If a particular heritage is to be preserved, the implication is that meaningful observations counter to the reigning paradigm should not be permitted. There is cultural continuity in the stabilized bits of knowledge associated with identity and heritage. It is what is known for sure, what is a given fact. What are considered socially stabilized facts "are experienced as being independent of opinion, belief, and cultural background."[7] The difficulty for both First American science and American Indians is that if it is perceived that facts are being preserved in the face of considerable countering information—such as was the case of Clovis-first for scientists or Kennewick Man for American Indians— then in popular American culture it loses its legitimacy and, therefore, loses its privileged position vis-à-vis dominant American cultural beliefs. The net effect of this erosion in the epistemological status of science resulted in actions such as congressional passage of the Native American Graves Protection and Repatriation Act (NAGPRA), which legally equates the conclusions of science and American Indian oral history for the control of certain human remains. In these instances, science is no longer privileged to determine its own set of facts. The political process, increasingly the courts, makes the ultimate determination.

In circumstances where scientific theories have the appearance of being value laden, there is a "prospect of a science whose results are continually in contestation."[8] When a theory such as Clovis-first is granted a status that allows preemptive dismissal of factual challenges—indeed, when the *theory* is revered as a fact rather than only a hypothesis with explanatory power for a particular set of observations—it "creates a bond between evidence and hypothesis impossible to break and even destroys ultimately, the concept of evidence as something to which one can appeal in defending a hypothesis."[9] While granting a theory the status normally accorded a fact might in the short run strengthen the prestige of the theory, it ultimately weakens the entire basis of scientific methodology; once a theory is successfully challenged, the previously sacrosanct facts can now also be challenged as being no better than a theory. What was a fact has now become conjecture.

There is an interesting heritage-related phenomenon that occurs in scholarship dealing with First Americans that generally does not occur elsewhere. On occasion, the tribal affiliation of the author is prominently mentioned as if it is an authoritative credential in a text that is produced for the academic community. For example, in *Opening Archaeology: Repatriation's Impact on Contemporary Research and Practice*, edited by Thomas W. Killion, there is a list of contributors that included the organizational affiliation of each chapter author. Examples are "Department of Anthropology, Wayne State University," "School of Human

Evolution and Social Change, Arizona State University," "Repatriation Office, National Museum of Natural History," and "Hanford Cultural Resources Laboratory, Pacific Northwest National Laboratory," all of which are certainly of sufficient stature to lend some aura of authority to a contributor's offering.[10] While there was no other biographical information—such as academic degree or publication history—provided in support of a contributor's credentials, two of the authors were further identified with the parenthetical "(Choctaw)" after their names.[11] In other times and places (e.g. the Jewish science of Hitler and Stalin), such an ethnic designation would have been considered not only irrelevant, but also fundamentally racist. Is it pertinent as to whether a particular scientific conclusion is generated by a self-identified *modern* American Indian? If it is not, then the tribal affiliation of an author is not only irrelevant, but also may lead the reader to question whether there is some other epistemological basis for the scholarship that is outside of, and inappropriately being expressed in the terms of, science. If, however, the tribal affiliation is considered appropriate, then the challenge is to determine the basis for such a conclusion. If a First American scientist is to be acknowledged as an American Indian, there must also be an understanding as to why there is no identification prominently emphasized for a current German political scientist writing on the politics of the Weimar Republic, for a Bantu mineralogist analyzing the Witwatersrand Gold Rush in South Africa, or for a modern Ethiopian paleontologist investigating human origins in the Great Rift Valley.

To the extent that it might be considered acceptable for authors in scholarly texts to publicly identify themselves as members of the very group that is being investigated, recognition is due, perhaps, to the work of feminist scholars. Tamara Bray, in addressing the role of power in First American science, drew on the theories of situated knowledge and standpoint epistemology of Donna Haraway and Sandra Harding, respectively, to situate First American theories. Many First American scientists, however, may not agree with the feminist conclusion that a complicity exists between cultural priorities and scientific methodologies. This alleged complicity is the source of the tension created by standpoint epistemology and situated knowledge, and the seeming oddity of academic authors listing their tribal affiliation in journal articles. Francis P. McManamon, then the chief archeologist of the National Park Service, concluded in 1999 that "public disagreements among scientists concerning methods, techniques, the appropriateness of some kinds of research, access to data, and other topics suggest to some nonscientists that the scientific enterprise has serious flaws and is untrustworthy."[12] While an author-associated tribal affiliation may prove to be of no consequence for conclusions reached in any particular scholarly effort, such an affiliation can be perceived as an open challenge to the mainstream academic system as a whole.

Given the dominant American *episteme* of scientific rationality, it is perhaps easier to isolate instances of active heritage preservation efforts by American Indians than it is to locate readily examples of unwarranted protection of theories (i.e., heritage preservation) by scientists. Theory selection is influenced or even determined by cognitive values, and every person will not come to the same decision in theory selection. Moreover, any selected set of cognitive values may well be inherently contradictory. It is not difficult to imagine instances when different approaches might be taken depending on the relative conceptual weight given to theoretical simplicity rather than accuracy. Yet there are instances in First American science—the case of the Clovis-first model is a good example— where defense of a particular paradigm reflected an almost religious devotion. According to the social psychologist Jonathan Haidt, "if a group circles around sacred values, they will evolve into a tribal-moral community.... They'll embrace science whenever it supports their sacred values, but they'll ditch it or distort it as soon as it threatens a sacred value."[13] As Mark Twain captured it, "you can't depend on your eyes when your imagination is out of focus."[14] If a theory is dependent on conservation of a particular set of values, then the theory may well be out of focus and impact the observations themselves. A particular view of the world is premised on a preserved set of values, a heritage. When a theory changes, the world changes too.[15] If a person, or even an entire culture, is comfortable with the world as it is perceived to exist, then there is a presumptive preservation of that world—preservation of its cultural theories, its heritage—whether by scientists or by an indigenous people.

For American Indians, the dominant Western cultural framework provides a pervasive challenge to their self-defined heritage. Within a hegemonic value structure, there is "a community of assumptions," according to Robert Merton, "upon which the entire system is built by the utilization of reason and experience. Within each context there is rationality, though the bases are non-rational."[16] If the initial conditions on which a heritage is constructed remain unchallenged, the resulting system appears to be rational. Michel Foucault addressed the potential absurdity of blindly accepting a seemingly rational system where the assumptions—Merton's "appropriate cultural conditions"—remain unexamined. In *Madness and Civilization*, Foucault described a man who believed that he was made entirely of glass; his behavior reflected this belief. Given this man's initial assumption that he was made from glass, Foucault concluded that, given his concern with his fragility, his subsequent behavior was fully rational.[17] From this, one should not conclude that either American Indian or scientific values are irrational, but rather that any examination that is conducted from *within* a value-framework is likely to be self-supportive in its conclusions.

Examined as a heritage preservation process, late nineteenth-century anthropologists were concerned with, among other things, providing what they believed to be an acceptable scientific explanation that would morally ratify the existing racially stratified society of the time. The monogenists, in consonance with the biblical account, believed that there had been a single human origination, but that some parts of humanity had degenerated into races that were clearly distinguishable. Others believed in polygenism, meaning that the several races had originated separately with the even clearer implication of a ranking of not only races, but what might be termed species. Despite the racist overtones of some monogenists who believed in the deterioration of certain parts of humanity, it was even easier for polygenists to support slavery.[18] Popular Western heritage preservation required a hierarchical differentiation between the races, and the science of both monogenists and polygenists was able to support such a conclusion.

In the mid-1860s, the Anthropological Society of London sponsored a series of papers by some of the leading intellects of the day that were read before the society and subsequently published. Volume I, covering fourteen papers read during 1863 and 1864, addressed what were considered the more primitive societies and races. Titles included: "On the Negro's Place in Nature"; "On the Weight of the Brain in the Negro"; "Some Account of the Astronomy of the Red Man of the New World"; and "The Neanderthal Skull: Its Peculiar Conformation Explained Anatomically."[19] In a paper read on May 12, 1863, titled "Observations on the Past and Present Populations of the New World," William Bollaert (1807–76) presented his analysis of the pre-1492 New World population and the subsequent adverse impact—through disease, war, and integration—that the European arrival had had on the total population.[20] Yet, in what was, even by modern standards, an otherwise cogent demographic analysis, Bollaert embedded polygenism. Although he first acknowledged "the red or copper-coloured men of the New World to be of one species of the genus *Homo*," he caveated it with the conviction that they were "of a different physiological condition when compared with the other species of mankind." He believed that "if we lean to the polygenistic view, or that of separate creations, it will, I think, render our study of Anthropology and Ethnology less complex."[21] As evidenced by what Bollaert believed to be legitimized science, cultures have a heritage that is embraced by scientists and embedded, unconsciously or otherwise, in the very science that they are conducting. However, when a heritage is historicized—that is placed in a retrospective context according to a different set of social values that are external to the regime itself—it may well be redesignated as myth rather than science, as has been the case with polygenism and other race-related aspects of the social sciences.

Deloria related a personal experience associated with court cases subsequent to the 1973 confrontation between US government law enforcement agencies and American Indian political activists at the Pine Ridge Reservation in South Dakota. According to Deloria, an observer of the court proceedings, which had discussed the Bering Strait theory, commented to him, "Well, dearie, we are all immigrants from somewhere." Deloria recalled thinking that "yes, indeed, but it makes one helluva difference whether we came 100,000 years ago or just out of boat steerage a generation back."[22] There may be an intuitive empathy with Deloria's point, but it is clear that he and his "observer" are talking past each other's value systems; they are situated in different heritages. The heritage embedded in First American science can easily conclude that: (1) no one, including any possible ancestors to modern American Indians, was here one hundred thousand years ago, and it is not even clear that anyone was here only fifteen thousand years ago; (2) modern American Indians may well not be descended from the First Americans, so no one here today is a true First American immigrant of some sort; and (3) no "helluva difference" actually exists in the New World when compared with the massive volume of Old World human migrations and conquests over tens of thousands of years. Counterpoised against this is Deloria's American Indian heritage that confidently posits a continuous occupation for the New World since the beginning of time.

Science studies has provided an extensive set of literature describing the confrontations between science and what has been termed lay expertise.[23] Although the terms employed are different, echoes of the same issues resound throughout the search for the First Americans. In a 1967 article titled "Scientific Colonialism," Johan Galtung characterized the anthropological equivalent of lay expertise as that of practicing a science as if the gravitational center of knowledge about a culture is situated someplace outside of that culture itself.[24] In the United States, legislation such as NAGPRA has provided a political intervention that has forced a shift in that epistemological center of gravity.[25]

Contentious issues of First American heritage and identity are not confined to the nexus between scientists and American Indians. Conflicts have also arisen between countries amid accusations of nationalism, colonialism, and imperialism. As the archaeologist Randall McGuire has commented, "One problem that nationalist movements encounter in creating national histories and heritages is that the past is already taken. . . . One group's defeat is another's victory."[26] As a science, American archaeology has been accused of being naïve in not acknowledging its historic relationship with the politics of colonialism and imperialism,[27] and of using the "past to legitimate the present . . . including depictions of the past in museums."[28] Nations tend to accept archaeological interpretations that support "how the people of the nation acquired their rightful territory or how they defended it from usurpers."[29] The extreme example of the use of archaeology to further nationalist goals was the "German archaeologists [who] spread across Eastern Europe looking for Germanic sites that would demonstrate Aryan racial superiority and justify the expansion of the Third Reich to include all of ancient Germany."[30] In the United States, First American science, particularly archaeology, has been "seen as a critical part of America's national identity, as it documented the country's progression from 'savagery' to the most 'civilized' place on earth. In 1906 this heritage was formally entrusted to science" when Congress passed the Antiquities Act.[31]

The once long-standing Clovis model held that Clovis Man—employing the then-advanced technology of a fluted stone spear tip—entered the New World through Beringia and in a matter of only one thousand years swept all the way to the tip of South America. In support of this theory, articles have been published in American academic journals that have concluded that based on "the culmination of morphological changes . . . a strong historical connection between North American Clovis points and South American 'Fishtail' points can be supported."[32] The authors, however, reached this conclusion even though the same article found that "South American forms would seem to have little in common with North American fluted points," that "all of these chronological data is far from clear-cut," and that the data "do not refute the idea that fluted point technology developed in the interior of western North America and . . . spread . . . eventually to the south into Central and South America."[33] The reader is left with the clear impression that the Clovis model was so powerful in the authors' minds that the absence of data refuting it was tantamount to a confirmation of the theory, and that the absence of similarity in the spear points resulted from morphological changes that would surely have confirmed the Clovis-Fishtail connection if only there had been any data. Others, such as David Meltzer, have concluded that Clovis artifacts have "never been found" in South America.[34]

For the five-hundred-year history of what is today the United States, David Hurst Thomas concluded that three distinct national narratives have developed.

First is the prominent glorification of America as a melting pot assimilating diverse cultures. Second is the American view that as a part of scholarly examination all subjects, including both American Indians and any artifacts potentially associated with First Americans, should be available for scientific inquiry. Finally, Americans believe that there is a perspective maintained in the oral traditions of the indigenous American population that is worth preserving. The proponents of each narrative are, of course, convinced that the American heritage that they espouse is the legitimate version "that should be published in textbooks, protected by law, and defended in the courtroom."[35]

The practical application of American Indian heritage preservation has translated into what is now called Cultural Resource Management, or CRM, for both First American and other American Indian artifacts. According to Randall McGuire, heritage has been translated into a nostalgia for the past (which may or may not have actually existed), which is packaged and sold as a commodity.[36] Heritage is to be preserved, culture is to be managed, and both are to be marketed for financial and political advantage. With the rapid increase in funding available for CRM, much of archaeology has been transformed from a "relatively limited, primarily research-oriented activity done by universities and museums to a multimillion-dollar, preservation-oriented business."[37] Replacing universities and museums are for-profit companies that are awarded contracts in which "the results of archaeological research are commodified as a 'deliverable' produced" at a competitive price.[38] Darby C. Stapp—an anthropologist who has not only written extensively on CRM, but has also been professionally employed on CRM contracts through the Battelle Memorial Institute—believed that "the result is that cultural resource professionals are now less responsive to the discipline of American archaeology and more responsive to the protection and management requirements of CRM and the interests of their employers, whether tribes or agencies."[39] According to the anthropologist David Killick and the archaeologist Paul Goldberg, there are "many more" archaeologists employed through CRM contracts than by universities and museums; moreover, "Most CRM reports do not undergo rigorous peer review."[40] Compliance archaeology, which provides the major portion of the funding that flows through CRM, results from requirements dictated by federal legislation such as NAGPRA. Joe Watkins—an archaeologist and a self-identified Choctaw—has concluded that academic archaeology "is in nearly direct conflict" with the business and government-based archaeology of CRM.[41]

The internecine struggle in the United States to control an important element of its existential myth—the cultural heritage associated with the First

Americans—is largely absent in the rest of the world. The archaeologist Geoffrey Clark believed that academic archaeology is better financed in the rest of the developed world than it is in the United States. Elsewhere, it is "perceived—rightly or wrongly—to be archaeology of the indigenous inhabitants, rather than the archaeology of the 'other,' as is the case in North America."[42]

CHAPTER 11

# Native American Graves Protection and Repatriation Act

B
EGINNING IN ENLIGHTENMENT EUROPE and continuing even into
the challenges of postmodernism, Western governments have perceived
science as providing a superior guide to the physical world. Moreover, it
has frequently been a metric against which other domains, including religion,
have been measured. As Jace Weaver, a professor of religion as well as law and
also director of the Institute of Native American Studies at the University of
Georgia, observed, "one can sympathize with Vine Deloria Jr.'s plea: 'science
should drop the pretense of absolute authority with regard to human origins
and begin looking for some other kind of explanation that would include the
traditions and memories of non-Western peoples.'"[1] The anthropologist Larry
Zimmerman echoed Deloria when he stated the need for "a different kind of
science between the boundaries of Western ways of knowing and Indian ways
of knowing."[2] As early as 1898, Jeremiah Curtin captured what has become the
essential tension between the beliefs inherent in American Indian myths versus
the scientific search for material solutions: "Of course there is no true information in the American ethnic religion as to the real changes which affected
the world around us; but there is in it, as in all systems like it, true information regarding the history of the human mind. Every ethnic religion gives us
documentary evidence."[3] For Curtin as well as for many later First American
scientists, American Indian myths have only a little to say about the physical
world around us.

As an exhibit at the Smithsonian Institution National Museum of the American Indian (NMIA) pointed out, possibly as much as 90 percent of the indigenous North American population was eradicated beginning with the fifteenth
century arrival of Europeans. What is not clear, however, is the extent to which
any ancestors of the surviving 10 percent were among those who were the First
Americans. In that same exhibit, Paul Chaat Smith, an associate curator at the
NMAI and a self-identified Comanche, stated that the surviving 10 percent
were "responsible for remembering everything, especially those things we never

knew."[4] This gets to the heart of the contest between science and American Indians in the search for the First Americans. For First American scientists, it is conceivable that modern American Indians are descended from later migrations to the New World and may not be descended from the First Americans; early Americans yes, earliest Americans, possibly no. Conversely, for American Indians, some believe that they are not only responsible for remembering everything, but also for ensuring that those memories remain intact; because their traditions, their myths, provide them an existential memory of continuity, they are confident they are descended from the First Americans.

Given the culturally sensitive nature of the potential answers associated with the search for the First Americans, both the questions asked and the methods used have become inextricably painted with some measure of sensitivity as well. Consequently, the American political process has been used as a lever to challenge the primacy of science. With regard to the evidentiary status of certain artifactual remains, the Congress, demonstrating that knowledge and power are coproductions, has declared that American Indian folklore and oral traditions are to be accorded equal status with science for the disposition of certain ancient human remains. In *Leviathan and the Air-Pump*, as a part of the philosophical struggle between Robert Boyle and Thomas Hobbes, Steven Shapin and Simon Schaffer characterized the correlation between approaches to nature and approaches to social order. As Jan Golinski concluded, *Leviathan* "began with disputes over 'technical' facts and argued outward to the broader issues that were revealed to be at stake, rather than arguing from the social context inward to technical content."[5] However, in the contested search for the First Americans, the flow is assuredly inward, from the social to the technical, from the right to define the existential inward to the ability to perform genetic analysis.

In a reversal of the traditional perception of science defining truth for political leaders, Congress has used power to referee who gets to assert truth. On November 16, 1990, the Native American Graves Protection and Repatriation Act (Public Law 101-601; 25 U.S.C. 3001 et seq.), commonly referred to by its acronym of NAGPRA, was signed into law and mandated a kind of epistemological leveling with regard to certain ancient human remains located in the United States. The stated purpose of NAGPRA as explained in the House of Representatives report that accompanied the legislation is to: "protect Native American burial sites and the removal of human remains, funerary objects, sacred objects of cultural patrimony on Federal, Indian and Native Hawaiian lands. The Act also sets up a process by which Federal agencies and museums receiving federal funds will inventory holdings of such remains and objects and work with appropriate Indian Tribes and Native Hawaiian organizations to reach agreement on repatriation or other disposition of these remains and objects."[6]

Science is no longer to be accorded a position superior to that of folklore and oral tradition in determining the disposition of certain artifacts. "An implicit consequence of repatriation legislation is the negation of Americanist archaeology's claims to exclusive control over the interpretation of the past. The alternative epistemologies and systems of knowledge authorized by the legislation challenge the privileged position of traditional academic knowledge."[7] In a statement that would surely be endorsed by American Indian organizations, Paul Feyerabend contended that "knowledge is obtained from a multiplicity of views . . . [and] that proliferation may have to be enforced by non-scientific agencies whose power is sufficient to overcome the most powerful scientific institutions."[8] Congress, certainly at once a powerful and nonscientific agency, has dictated just such a multiplicity of views in this instance.

In determining the cultural affiliation of human remains and funerary objects found on federal property or located in an institution supported with federal funding, US federal law now holds that a determination of culture affiliation shall rest on "a preponderance of the evidence based upon geographical, kinship, biological, archaeological, anthropological, linguistic, folkloric, oral traditional, historical, or other relevant information or expert opinion."[9] Frequently, identification of the latter category, expert opinion, is more difficult than might be imagined. The Nobel Laureate physicist Richard P. Feynman once commented that "science is the belief in the ignorance of experts."[10] Although Congress has not declared scientists to be ignorant, it has declared in law that expertise can, and does, exist outside of science.

The ramification for potential archaeological sites is enormous. More than a quarter of the US land mass is owned by the federal government and, therefore, is subject to the constraints of NAGPRA. In addition, NAGPRA also covers archeological work and institutions receiving federal funds. Finally, lands controlled by American Indian organizations will presumably be at least as restrictive for excavation and as any limitations that NAGPRA might dictate.

This legal edict of epistemological equality in the search for the First Americans may have succeeded only in shifting the venue of the theoretical arguments from archaeological sites and genetic laboratories to litigious America's courtrooms—as was evident with the remains of the Kennewick Man, which is discussed in Chapter 12. The American political process has legislatively mandated an equality between folklore and oral tradition, on the one hand, and biology, archaeology, and physics, on the other, with regard to certain Native American human remains and funerary objects. Science must be able to take on all opponents in its search to characterize nature. However, the political leveling is not evenhanded in endorsing nonscientifically based ontologies. The treatment NAGPRA grants American Indian traditions contradicts the position taken by US courts in consistently according science a privileged position vis-à-vis Christianity—which could easily be equated with folkloric and traditional beliefs—with regard to evolution and other matters. Clearly, American Indian traditions have been endowed with a privileged legal status vis-à-vis science when compared with other religions and cultural traditions.

When NAGPRA was signed into law in 1990, *Archaeology* noted that 623 US federally funded museums and research institutions had 157,000 bodies and millions of funerary objects that were believed to be associated with indigenous American Indian and Native Hawaiian cultures.[11] Jace Weaver in *Wicazo Sa Review*, citing the work of Walter and Roger Echo-Hawk, estimated that "museums, federal agencies, other institutions, and private collectors retain[ed] between 300,000 and 2.5 million dead bodies taken from Indian graves, battlefields, and POW camps by soldiers, museum collectors, scientists, and pothunters."[12] Determining the precise numbers is less important than recognizing that they are large, and that over the following twenty years, "only a quarter of all the human remains have been culturally affiliated or repatriated."[13] The symbolic value of congressional action was extremely important both to scientists and to American Indians. According to Thomas W. Killion, NAGPRA is an important social and political milestone for American Indians. It involves "issues of cultural survival, community revitalization, knowledge and language preservation, protection of sacred sites, and political sovereignty."[14] It plays "a powerful and pervasive role in the development of the discipline" of American archaeology.[15] Advances in First American science are "unrelated to the 'science as usual' routine thought to prevail in archaeology. . . . Rather, this potential for an intellectual change of perspectives is externally driven."[16]

Although NAGPRA is the current legal centerpiece for regulating the relationships across First American scientists, Indian organizations, federal agencies, and museums, it has not been the only such effort by Congress to legislate the procedures of First American science. The Antiquities Act (AA) of 1906 (Public Law 59-209; 16 United States Code 431-433) in four short sections makes no mention of American Indians. Moreover, it grants power to federal executive agencies to ensure that the "examinations, excavations, and gatherings [of antiquities] are undertaken for the benefit of reputable museums, universities, colleges, or other recognized scientific or educational institutions, with a view to increasing the knowledge of such objects, and that the gatherings shall be made for permanent preservation in public museums."[17] In 1906, First American science was officially assured of epistemological primacy by Congress.

Another significant congressional action was the Archaeological Resources Protection Act (ARPA) of 1979 (Public Law 96-95; 16 United States code 470aa-mm), which even today mandates a requirement to obtain a permit prior to archaeological excavation on federally controlled lands. While still providing science a privileged position, it is clear that by 1979 the evolution to what became NAGPRA was already underway. The emphasis in ARPA by Congress was less about the role of science than it was a reflection of the dominant American culture's interest in the search for the First Americans. For example, Section 2(a) (1) of the law stated that "archaeological resources on public lands and Indian lands are an accessible and irreplaceable part of the Nation's heritage." Later in Section 2(b), the act stated its purpose as being "to secure, for the present and future benefit of the American people." However, the role of First American science was protected in Section 4(b)3 through a legal mandate that "the archaeological resources which are excavated or removed from public lands will remain the property of the United States, and such resources and copies of associated archaeological records and data will be preserved by a suitable university, museum, or other scientific or educational institution."[18] There is no mention of repatriation of the human remains to American Indian tribes, nor even an acknowledged responsibility incumbent upon First American scientists to consider the traditions of modern American Indians.

Despite the clarity of Congress's intent in enacting NAGPRA, the power relationships between First American scientists and American Indian tribes is far from stable. Amended by the Congress and further interpreted by regulations issued by the secretary of the interior, there are continuing efforts from all sides to modify NAGPRA. In support of American Indians, S. 2843 was introduced in 2004 by US Senator Ben Nighthorse Campbell of Colorado; it would have altered the definition of Native American cultural affiliation by modifying "is indigenous to the United States" to make it "is or was indigenous to the United States" in the existing NAGPRA law to permit modern American Indians to associate themselves more easily with the remains of ancient indigenous people.[19] Campbell's effort to modify NAGPRA was a direct result of the decision in the Kennewick court case, which favored the scientists' position over that of the American Indian tribes. In a 2007 example in support of scientists, Washington Congressman Doc Hastings introduced H.R. 4027, which would require that human remains found on federal lands must be demonstrated to have a *significant* relationship with existing tribes before they can be repatriated.[20] Similar efforts to amend the scope of NAGPRA continue through the legislative and regulatory processes.

Under the authority granted to the secretary of the interior in the NAGPRA legislation, in 2010 the Department of the Interior modified the extant rules to address the status of the remaining 115,000 culturally unidentifiable human remains. Before requiring repatriation, the original law required that a *cultural* affiliation be established between a modern American Indian tribe and the human remains. Now, the regulation requires the establishment of only a *geographical* connection—a tribal claim to a historic territorial habitation—as sufficient to establish a claim to repatriation without requiring the demonstration of cultural affiliation as indicated in the original NAGPRA law.[21] Citing critical letters to the secretary of the interior from the National Academy of Sciences, the Society for American Archaeology, and the American Association of Museums, *Archaeology* cast the change as another step in the erosion of scientific authority.[22]

Given the history of desecration of American Indian graves, it is difficult to challenge the moral basis of NAGPRA. In 1979, the Department of the Interior concluded that "the prevalent view in the society of applicable disciplines is that Native American human remains are public property and artifacts for study, display, and cultural investment. It is understandable that this view is in conflict with and repugnant to those Native people whose ancestors and near relatives are considered the property at issue."[23] The legislative history documents unsavory instances such as when the surgeon general in 1868 ordered all Army field officers to send to him American Indian skeletons for study so that he could determine whether the American Indian was inferior to the white man based on the size of the cranium.[24]

In a July 17, 1990, hearing before the House Committee on Interior and Insular Affairs, those representing American Indian interests testified that the scientific community quickly analyzed and reburied non-Indian remains while Indian remains were frequently sent to museums. Scientists countered with the importance of using human remains for scientific study and that wholesale reburial of Indian remains would preclude further scientific investigation as scientific techniques are improved. Finally, "witnesses representing private art dealers testified that Native Americans should not be the sole conservators of their cultural items because all Americans have a right to their history."[25] In the case of the art dealers, "ownership" of the past is evidently to be taken literally.

The degree to which science is embedded in the modern American *episteme* becomes clear in these debates over human remains. American Indian activists have concluded that "First Amendment religious freedoms are clearly controlled from the pulpit of science when museums elevate scientific curiosity over Indian religious belief in the treatment of the dead."[26] Interestingly, those who are resisting the reach of scientists appear also to espouse the notion that there is a single science that has a rationally coordinated agenda.

The conflict between First American scientists and American Indian organizations over human remains reached a contentious zenith with the Kennewick Man. However, there have been many other disputes. One conflict centered on a lock of human hair, involving on one side the Confederated Salish-Kootenai and the Shoshone-Bannock tribes, and on the other side the archaeologist Robson Bonnichsen, a key participant in the Kennewick affair and the founder of the Center for First Americans located initially at Oregon State University and currently at Texas A & M University. Bonnichsen discovered human hair at a ten-thousand-year-old archaeological site in Montana and planned to subject it to DNA analysis. The two Indian organizations petitioned the federal Bureau of Land Management, which controlled the location, to bar Bonnichsen from the site, to prohibit the tests, and to require the repatriation of the hair to the tribes. According to the Smithsonian Institution's Douglas Owsley, "the application of NAGPRA legislation has gone beyond its original purpose, which was primarily concerned with affiliated remains of more recent context."[27] After two years, Bonnichsen prevailed and NAGPRA regulations were amended to exclude naturally shed human hairs from the repatriation provisions of NAGPRA.[28]

There appears to be little consensus over the degree to which NAGPRA actually impedes scientific study. For example, Francis P. McManamon believed that there is no adverse impact. "Although it does constrain some kinds of scientific research, NAGPRA does not prohibit research on Native American human remains, funerary objects, or other cultural items covered by the law. In fact, effective implementation of the law requires some kinds of research. For example, in order to determine the appropriate cultural affiliation of certain Native American human remains, scientific investigation, analysis, and interpretation of certain kinds of data are required."[29] However, the anthropologist Keith W. Kintigh countered that governmental agencies frequently succumb to political pressure from American Indian organizations. It is Kintigh's contention that despite the requirement in NAGPRA for science to be given equal consideration with American Indian oral traditions, that government agencies "often accede to tribal demands that no drawings or photographs of burials be made, and they fail to require even the most basic identifications and assessments relevant to cultural affiliation." The result, according to Kintigh, is that the human

remains and artifacts are quickly reburied without documented cultural affiliation and without scientific examination, especially in cases where the most vocal
and most "stridently antiscientific" American Indian tribes are involved.[30] This
contention would seem to be borne out based on Judge John Jelderk's finding in
the Kennewick Man case (*Bonnichsen et al. v. United States et al.*) where he concluded that the US Army Corps of Engineers repeatedly and secretly gave American Indian tribal officials access to preliminary information that was denied to
the interested First American scientists, and that the government's behavior was
consistently "marked by an appearance of bias."[31] Thomas W. Killion, a professor
of archaeology at Wayne State University, however, countered that those who
believe that there is an "assault-on-science" that threatens scientific practice by
favoring political considerations over the "disinterested evaluation of knowledge
claims," which are not recognizing "all knowledge claims, to a lesser or greater
extent, are intimately bound up in considerations that are inherently 'political'
and otherwise 'interested.'"[32] The debate is far from resolved even within the
First American science community.

Neither First American scientists nor American Indian traditionalists have
developed a monolithic position. Archaeologists frequently insist they have a
privileged position in representing First American science in the search for the
First Americans. Similarly, some American Indian tribes are concerned that
NAGPRA must not adversely affect their preferential power relationships with
the federal government vis-à-vis other tribes. According to Joe Watkins, as a
result of NAGPRA, federally recognized tribes "were afraid that a federal agency
could grant an implied status of 'Indian tribe' to a non-federally recognized Indian tribe by 'recognizing' the tribe as eligible to participate in the repatriation
process; they were afraid that the recognition of the tribe as eligible for the special programs and services of a federal agency might allow the tribal group to
apply its eligibility to other programs and become 'federally recognized' without
undergoing the rigorous BIA [Bureau of Indian Affairs] recognition process."[33]
In other words, federally *recognized* tribes wanted to exclude *non*-recognized
tribes from enjoying the same position that they occupy—that is, being
self-determinant in their tribalness. Oddly, these tribes were relying for their
recognition on an external (i.e., a science-centric and political approach by the
federal government) rather than the internal myth-centric self-determination
privilege that these same tribes demanded from scientists and the federal government in claiming their human remains. That American Indian tribes are aware
of the importance of the distinction made by the federal government was shown
in a Smithsonian National Museum of the American Indian exhibit by curators

of the Pamunkey Tribe of Virginia. As indicated, "unlike many other tribes, the Pamunkey have yet to be recognized as a sovereign nation by the United States." Importantly, "tribal members differ on the value of this recognition."[34]

Is NAGPRA a blatant political intervention in science? If the *New York Times* can be considered a reflection of the dominant ethos, a prism into Foucault's *episteme* of modern American culture, then the popular perception is indeed that NAGPRA has permitted a political and detrimental intervention in the scientific process. On October 22, 1996, the *Times* titled an article on the Kennewick Man "Indian Tribes' Creationists Thwart Archeologists."[35] Jace Weaver described the article as "heavily slanted in favor of the scientists, [and] likened Native Americans seeking return of ancestral remains pursuant to the provisions of NAGPRA to Christian fundamentalists."[36] According to a *Times* excerpt cited by Weaver: "Since the repatriation act was passed in 1990, American Indian creationism, which rejects the theory of evolution and other scientific explanations of human origins in favor of the Indians' own religious beliefs, has been steadily gaining in political momentum. Adhering to their own creation accounts as adamantly as biblical creationists adhere to the book of Genesis, Indian tribes have stopped important archeological research on hundreds of prehistoric remains."[37]

Both science and religion command cultural allegiance from their adherents. In *Leonardo's Mountain of Clams and the Diet of Worms*, Stephan Jay Gould (1941–2002) introduced the concept of Non-Overlapping Magisteria, NOMA for short, after a visit to the Vatican's Academy of Science in Rome.[38] As a part of NOMA, Gould contended that the jurisdictions of the religious and scientific domains do not overlap and that the frequent war-like metaphors used to describe the relationship between science and religion are not appropriate. As appealing as Gould's supposition might be, it is not borne out by the intensity of the charges and countercharges in the First American debate. According to the archaeologist Robson Bonnichsen, "this is a battle over who controls America's past."[39] Deloria contended that the theory that the Americas were populated during the last ice age by migrations from Asia was "scientific folklore."[40] The existential myths of many American Indian tribes asserts that they have been in their modern homeland since their creation. The fear—as Jace Weaver characterized it in the context of the ten-thousand-year-old Kennewick Man's reported European cranial characteristics—is that "the Bering Strait theory, having been used to make Indians immigrants no different fundamentally from those who disembarked at Ellis Island, is now used to make them European as well, to make the populating of the hemisphere no different than Columbus's (or the Vikings') 'discovery' centuries later."[41]

That politics is involved in the search for the First Americans, whether from the scientists or from the American Indians, is obvious. This intervention may be generated by a plainly evident political agenda or may originate from values more subtly embedded in a group's particular cultural ethos. Randall McGuire stated that First American scientists *can* "use their craft to evaluate interpretations of the real world," but also *should* "construct meaningful histories for communities, to strive for real collaboration with communities and to challenge both the legacies of colonialism and the omnipresent class struggles of the modern world." He alleged, however, that this can only happen if First American scientists "surrender significant control over our research agenda."[42] If NAGPRA has not required that First American scientists cede control of the research agenda, it has surely mandated that this agenda is now subject to negotiation with communities external to their own. For American Indian tribes, NAGPRA is a mechanism for them to receive "culturally appropriate" consideration in their effort to ensure that their culture survives.[43] For some First American scientists, NAGPRA would appear to have redefined their role to providing a life-support system for Indian cultures rather than conducting an indifferent science.

# CHAPTER 12

# Case Studies: Kennewick Man and Tarim Basin

T HE EXPRESSION OF HERITAGE, identity, race, nationalism, and political power are all evident not only in the search for the First Americans but also in the search for (or suppression of) evidence of earlier cultures in other countries. This chapter first examines the controversial discovery in the United States of the remains of what has become known as the Kennewick Man, and then examines efforts by the Chinese government to suppress evidence of early non-Asian cultures in what is now modern China.

## Kennewick Man (The Ancient One)

In July 1996, two college students attending hydroplane races on the Columbia River near Kennewick, Washington, found a human skull. They called the police, who contacted Floyd Johnson, the local coroner for Benton County. Johnson in turn called James Chatters, a PhD archaeologist and paleoecologist who served as a consulting deputy coroner when his expertise was required. Upon examining the skull, Chatters "was immediately struck by its long, narrow shape and the marked constriction of the forehead behind a well-developed brow ridge. The bridge of the nose was very high and prominent. My first thought was that this skull belonged to someone of European descent." In addition to the shape of the skull, Chatters observed that "the nose had been huge," and that the jaw displayed "deep depressions behind the ridges formed by prominent canine teeth. Called a *canine fossa*, this is an archaic characteristic common to many European skulls." Based on these preliminary findings, Chatters believed that "this was a white person, a Caucasian."[1] What prevented this from being a routine police investigation was that, based on the dietary wear patterns on the teeth, Chatters also concluded that the skull could potentially be five thousand years old.[2] However, artifacts also recovered from the same site opened the much less interesting possibility that the remains were simply those of a several-hundred-year-old European pioneer.[3]

Over the next few weeks, the sequential headlines of the local *Tri-City Herald* captured not only the quickly evolving story, but also the American imagination: "Skull Found on Shore of Columbia," July 29, 1996; "Skull Likely Early White Settler," July 30, 1996; and "Tri-City Skeleton Dated at 9,000 Years Old," August 28, 1996.[4] After examination, Chatters concluded that the remains were approximately 9,500 years old and "bore no resemblance to the aboriginal peoples of the Columbia River basin."[5] Anatomically, Chatters described the Kennewick Man as being a male, approximately five feet, nine inches tall and having "a long, narrow skull with a narrow face and receding cheekbones" that was quite different from "the short, round skulls with broad, flat faces" that are characteristic of the modern America Indians of that geographic area.[6] The acrimony surrounding the human remains was evident in a 1997 article in the *Santa Fe New Mexican* by reporter Keith Easthouse, who asked "when Columbus came to the New World in 1492 and set in motion the chain of events that led to the decimation of Native Americans, was he unknowingly getting revenge for what was done to his ancestors thousands of years before?"[7] The incendiary combination of the skull being characterized as white and ancient was to ignite a rancorous and continuous debate across First American scientists, American Indian tribes, government officials, and popular American culture.

Because the Columbia River bank, where the skeletal remains had been found, was under the control of the US Army Corps of Engineers, the provisions of the Archaeological Resources Protection Act of 1979 (ARPA) were triggered. According to ARPA, a federal permit was required for further searches for human and artifactual remains as well as for control of the existing remains for examination purposes. Chatters applied for—and on July 30, 1996, received—the necessary permit.[8]

The physical evidence surrounding the discovery was unusually plentiful, and the full suite of the capabilities of First American science could be applied. Only a few ancient human remains have been found in the Americas.[9] When the Kennewick Man discovery was made, there had been only two substantive finds in the northwest of the Western Hemisphere, an 11,700-year-old in Washington and a 12,800-year-old in Idaho. As a result, most evidence addressing the First Americans is not from actual humans themselves, but rather is based on the environmental context within which ancient humans were active—that is, animal remains as well as the stone and bone tools believed associated with Paleoamericans. From the reasonably complete Kennewick Man skeleton, especially the skull and jawbone, it was possible to not only examine the cranial shape, the patterns of wear on the teeth, and the genetic composition, but also perform carbon dating on the human tissue. Based on the arrowhead embedded in the Kennewick Man's hip, it was also possible to evaluate the cultural tool kit and look for associations with documented ancient as well as modern artifactual evidence. As an academically trained and experienced First American scientist, Chatters pursued his examination by also recruiting recognized and independent experts for these disciplines.

To perform the radiocarbon dating, Chatters sent a metacarpal fragment from the left hand to Donna Kirner at the University of California, Riverside, who dated the remains at 8,400 years, BP. Supported by these results, Chatters estimated the age of the Kennewick Man to be approximately 9,500 chronological years old.[10] Later studies conducted under the auspices of the National Park Service generally supported Chatters's conclusions.[11]

With the antiquity of the Kennewick Man confirmed, and the remains being found on federal lands, the provisions of the Native American Graves Protection and Repatriation Act (NAGPRA), enacted in 1990 by the US Congress, came to the fore. NAGPRA requires that geographical location, American Indian oral tradition, and science are to be considered equal factors in determining the cultural affiliation of such ancient human remains. The Confederated Tribes of the Umatilla Indian Reservation declared the Kennewick Man, whom they called the Ancient One, to be a tribal ancestor and demanded the repatriation of the remains so that he could be immediately reburied in an undisclosed location in accordance with their traditions.[12]

Under the terms of NAGPRA, there are provisions that permit human remains in a museum's collection to be retained for a limited time for scientific examination. However, the Kennewick Man's remains were not part of a museum collection. In this case, there was little pertinent artifactual or biological evidence that suggested any relationship to modern, local American Indians. That left consideration of oral history and geographical proximity as determining factors. The logic of many American Indian tribes is straightforward. Since their historical beliefs hold that all human inhabitants of the pre-Columbian New World were non-European, then any remains that are found that are not clearly post-1492 must, of necessity, be their ancestors. Armand Minthorn, the Umatilla tribe spokesman, declared, "Our oral history goes back 10,000 years. . . . We know how time began and how Indian people were created. They can say whatever they want, the scientists. They are being disrespectful."[13]

On September 17, 1996, Lieutenant Colonel Donald S. Curtis Jr., the US Army Corps of Engineers Commander of the Walla Walla District Office on whose land the Kennewick Man had been discovered, announced in the local

*Tri-City Herald* that he intended to repatriate the remains to the five different tribes—the Umatilla, Yakama, Nez Percé, Wanapum, and Colville—that had demanded repatriation.[14] According to Francis P. McManamon, later named a defendant in the Kennewick Man court case, scientists need the support of the officials of federal agencies and museums if they are to conduct their research, but the same government officials must be familiar with the "scientific potential of the collections and sites they control. They must recognize the social benefits of research results that may be realized through scientific investigations."[15] Chatters, however, described McManamon as being "a career government employee with little practical archaeological experience."[16] According to David Meltzer, "the Corps works on hundreds of projects across the country, many on tribal lands, and it was not unnoticed that it was in their best interest to get along with the tribes."[17] Chatters also concluded that "if an archaeologist these days wants to continue working, he has to stay in the tribes' good graces."[18]

Based on previous instances of federal agency interpretation of the provisions of NAGPRA, there were real consequences for the practice of First American science as a result of NAGPRA. In 1991 human remains believed to be 12,800 years old had been repatriated to the Shoshone-Bannocks with time permitted for only an incomplete study by a graduate student. In 1993 a nine-thousand-year-old partial skeleton found in Hourglass Cave, Colorado, had been quickly repatriated.[19] There were not only ramifications for First American science if some of the American Indian tribes became irritated, but there could be personal consequences as well. According to Chatters, the Colville tribe wrote to the employer of each scientist involved in the Kennewick Man lawsuit asking if that institution wanted a reputation of having its employees trying to hurt American Indians. As a consequence of this campaign, Bonnichsen was charged by his department chairman at Oregon State University with ethics violations due to his being a plaintiff in the lawsuit.[20]

The government position, however, was not unified. On October 15, 1996, less than a month after the Army Corps of Engineers statement of intended repatriation, the Smithsonian Institution officially requested the remains on the basis that the 1846 act creating the Smithsonian Institution gave them jurisdiction over remains and artifacts recovered from federal lands.[21] Two of the eight plaintiffs suing for access by First American scientists to the Kennewick Man's remains were Douglas W. Owsley and Dennis J. Stanford, both at the Smithsonian Institutional National Museum of Natural History.

After the discovery of the Kennewick Man's remains, it required only two months before these churning confrontations were captured in the national press. On September 30, 1996, the *New York Times* published an article by Timothy Egan titled "Tribe Stops Study of Bones that Challenges Its History."[22] During the governmental thirty-day public comment period leading up to October 23, 1996, additional reports appeared in *Time, Newsweek, The Economist, Archaeology, Science, ABC, PBS*, and the *Discovery Channel*.[23] The tone in the media was that the contest not only was between science and American Indian religion, but also between Europe and Asia as the source for the First Americans.[24] Yet David Hurst Thomas, a First American anthropologist at the American Museum of Natural History in New York, believed that the "dispute is not a matter of religion v. science, or even Indians v. Scientists. At its heart, the matter of the Kennewick skeleton involves political power and property rights."[25]

The legal battle was officially joined in the US District Court when a group of First American scientists filed one lawsuit to allow a more thorough scientific examination, and the Asatru Folk Assembly filed another demanding reburial of the remains according to their religious doctrine. The basis of both lawsuits was Chatters's preliminary conclusion that the remains were Caucasian and, therefore, the provisions of the Archaeological Resources Protection Act of 1979 were controlling, rather than the provisions of Native American Graves Protections and Repatriation Act of 1990. In the lawsuit by the First American scientists—*Bonnichsen, et al. v. United States, et al.*—Judge John Jelderks of the US District Court in Oregon included in the court opinion that "The Plaintiff scientists are highly regarded experts in their fields," and then listed the professional employment of each of the eight scientists involved as plaintiffs. The issue of scientific expertise versus American Indian traditions has not only been central to the debate in the Kennewick Man case but has also pervaded the entire search for the First Americans.

While the American Indian tribes were not directly a party to the lawsuits, as active amici curiae participants, the court did consider the expertise of the American Indian tribes as mandated by NAGPRA. According to the court decision, the Confederated Tribes of the Umatilla had contended that "We already know our history. It is passed on to us through our elders and through our religious practices. . . . From our oral histories, we know that our people have been part of this land since the beginning of time. We do not believe that our people migrated here from another continent, as the scientists do."[26] However, this indigenous expertise was evidently less persuasive to the court than that provided by the First American scientists. According to Judge Jelderks's decision, "the issue is not whether Indian tribes are entitled to recover the remains and cultural objects of their own ancestors, but whether they also are entitled to claim remains and cultural objects having no demonstrated link to any present-day tribe or to modern American Indians in general."[27] Yet, according to the terms of NAGPRA, it would appear that the courts must weigh equally such evidence as might be provided variously by expertise situated in First American science, geography, and American Indian oral traditions.

Other groups had also come forward. The Asatru Folk Assembly, a religious movement of approximately five hundred people who claim to follow pre-Christian European Norse traditions that honor their ancestors,[28] filed a lawsuit that demanded repatriation of the Kennewick Man remains based on the Caucasian anatomical relationship.[29] According to Stephen McNallen, then the Asatru president, "We claim these bones of our ancient kinsman on the grounds that we are more closely related to him than are the Native American tribes in the area."[30] McNallen also commented that "We don't want to offend Native Americans," because "we have a lot in common with them."[31] Asatru, however, ultimately decided not to pursue the matter beyond filing of the initial lawsuit. The court directly considered the "Caucasian-ness" of the Kennewick Man skull. According to information considered by the court, an analysis of the characteristics of the skull indicated a significant probability of a relationship to the Polynesians and the Ainu of Japan rather than modern American Indians.[32] However, the court also noted that the analysis did not conclude that the Kennewick Man was Caucasian.[33] The Kennewick Man is not the only early human remains that has suggested a skeletal structure different from that of modern American Indians. Two specimens from Spirit Cave dating to ten thousand years old also have faces that are longer and narrower than that of American Indians.[34]

One issue that greatly concerned the First American scientists was proper preservation of the human remains, which had been seized by the Corps pending resolution of the lawsuit. As it turned out, the femurs from the Kennewick Man disappeared, and it was a year and a half before the Corps realized that the bones were missing, and nearly five years before they were located.[35]

Federally *recognized* tribes are given preferential treatment under NAGPRA and other US laws. American Indians from *non*-recognized tribes have considerably fewer culturally situated legal rights. The Ethnic Minority Council of America, one of the organizations interested in the Kennewick court proceedings, demanded that the remains be studied scientifically, and that reburial of unstudied remains was tantamount to "passive genocide" of the non-recognized tribes.[36] Interestingly, Chatters casts the Ethnic Minority Council as if it formally represented the interests of these non-federally recognized tribes, but its website made no such claim. Although the council indeed claimed to help protect the rights of American Indian tribes, the organization described its membership to be only individuals. There was little basis for Chatters concluding that it represented these particular tribes. There was, perhaps, a coincidence of interests, but a representational relationship is questionable. In a July 15, 2005, letter to Senator John McCain, the council claimed that "Many of our members are of American Indian descent and these members live both on U. S. Indian reservations and independently outside reservations."[37] However, it also acknowledged that the membership was much broader and included not only agnostics, but also Buddhists, Native Americans, Christians, Jews, Moslems, and Hindus. The politics of the search for the First Americans is fertile territory for the creation of surprising groupings of fellow travelers.

The already considerable difficulties associated with use of the term Caucasian to describe the Kennewick Man's skull were compounded when Chatters opined that a facial reconstruction of the skull produced a striking likeness to Patrick Stewart who had starred as Captain Jean-Luc Picard on the television series *Star Trek: The Next Generation*. The media began publishing side-by-side pictures of Stewart and the Kennewick Man reconstruction, which, unfortunately, did indeed bear some resemblance. However, others countered that the facially reconstructed Kennewick Man bore a strong resemblance to the 1833 John Jarvis painting of the American Indian chief Black Hawk and his son Whirling Thunder. As for Chatters, he believed that his comparison of the similarities between Stewart and Kennewick Man resulted in accusations that he was practicing racist science based on his use of what he called the "C-word."[38]

On June 27, 1997, Judge John Jelderks required the government to provide answers to seventeen detailed questions, including: Are the remains subject to NAGPRA? What does NAGPRA mean by the terms Native American and indigenous? Was there more than one wave of ancient migration to the Americas? If there were subpopulations of early Americans, does NAGPRA apply to skeletons from a people who died out and were not directly related to modern Indians? What happens to the remains if no existing tribe is culturally affiliated with them? Are scientific study and reburial mutually exclusive? Do the scientists have a First Amendment right to study the remains?[39] This echoes Adovasio asking, much less eloquently than Judge Jelderks, "Who the hell are these people, where did they come from, and when did they get here?"[40]

The rancor surrounding the disposition of the Kennewick Man's remains cannot be considered simply an arcane, but publicly stimulating, academic debate. As David Hurst Thomas concluded, the "matter raises formidable questions about tribal rights with regard to human remains and archaeological sites, treaty rights and Indian land claims, and especially the contemporary religious use of tribal and federal land."[41] A cottage industry has developed that follows the Kennewick Man controversy. The National Park Service even maintains a website that tracks the several academic studies and the legal confrontations that have occurred.[42]

On August 30, 2002, the US District Court in Oregon released its decision by Judge Jelderks, holding that the Kennewick Man's remains could not be considered to be American Indian, and that First American scientists were to be allowed access in order to conduct a detailed scientific analysis. On April 19, 2004, the Ninth Circuit Court of Appeals upheld the district court decision. Under the terms of NAGPRA, ownership of the remains continued to reside with the Army Corps of Engineers. However, the court ordered that the remains were to be kept physically under the "neutral" control of the Burke Museum of Natural History and Culture at the University of Washington.[43] As Meltzer concluded, "the claim the Americas were colonized by Caucasians has not gone away. Instead, it has seeped into the only place such foolishness passes as fact: the poisonous corners of the Internet where white supremacists continue to claim Kennewick as one of their own, to promote the idea that members of the Aryan race—whatever *that* is—were the real discoverers of America."[44]

The legal contest was finally resolved with the 2015 completion of a study by First American scientists that concluded that the remains were more closely related to modern American Indians than any other group. As a result, the remains of the Ancient One were turned over to the Confederated Tribes of the Umatilla and in 2017 were buried at an undisclosed tribal site near the Columbia River. In one sense, both First American scientists and the Confederated Tribes were winners in that the remains underwent an exhaustive scientific examination but were also buried by the tribes on their lands. In another sense, both were losers. The scientists no longer have access to the remains for examination as new analytical methods are developed, and the tribes suffered the indignity of having the remains of one of its most ancient ancestors desecrated to prove what they had contended in the first place. The interplay of power versus politics, science versus heritage was on full display in the contest between the Kennewick Man and the Ancient One.

## China's Tarim Basin

The contentious intersection of power, science, and cultural identity is not restricted to the democratic politics of the United States. Popularly characterized as a Communist dictatorship dominated by Han culture, China provides an excellent example of the same political forces experienced in the United States in the search for its First Americans.

The Xinjiang Uyghur Autonomous Region in western China is home to ten million Uyghur who speak a language related to Turkish not Mandarin; generally, practice Islam in the midst of an officially atheistic society; and have facial features typically resembling the people of Eastern Europe and central, not eastern, Asia. Moreover, the Uyghur themselves contend that they are not Han Chinese. While the cultural conflict between the ethnic Uyghur and the Han-dominated Chinese government is expressed at many levels, the use of science as a tool to leverage cultural identity arose in the context of ancient mummies—more properly desiccated corpses without any artificial preservation—found in the Xiaohe cemetery, first investigated by the Swedish archaeologist Folke Gergman in 1934, in Xinjiang China's Tarim Basin.[45]

Thousands of artifacts were discovered, including a two-thousand-year-old fabric resembling that found in an Austrian salt mine and a string-skirt undergarment typically associated with European textiles.[46] Although the occasional mummy discovery had originally been attributed to travelers along the ancient Silk Road, hundreds of similar ancient bodies, some mummified and others not, began being uncovered in the 1970s by construction workers. The Uyghurs, especially those advocating an independent Uyghur nation, cited these remains and their associated artifacts as evidence of ancestral claims to the lands of the Xinjiang region. For many years, the bodies disappeared, reportedly at the direction of the Chinese government.[47] There the matter rested, awaiting the development of modern genetics and China's more active engagement with the outside world.

While the popular press has focused on "the Beauty of Xiaohe"—an auburn-haired, European-featured, 3,800-year-old mummy—archeologists working in the Tarim Basin have uncovered hundreds of graves, some dating to four thousand years ago, with similar physical characteristics.[48] Other mummies have such colorful names as Charchan Man, Yingpan Man, the Witches of Subeshi, Shaman of Yanghai, and the Man from Hami with a Dozen Hats.[49] While the mummies may not be the direct ancestors of the Uyghurs, they are surely proof of a civilization in the region that predated the Han Chinese·

In the early 1990s, Victor Mair, a professor of Chinese language and literature at the University of Pennsylvania, and a team of primarily Western scholars were granted permission by the Chinese government to begin archaeological, linguistic, and genetic research on the mummies and artifacts of the Tarim Basin.[50] In addition to the European-associated fabric and the physical characteristics of the mummies, texts dating to 500–700 CE indicated the language of the time to be not of East Asian origin, but rather Tocharian, an ancient Indo-European language that probably separated from its European roots around 5,600 years ago.[51] It was genetic testing, however, that proved to be at the heart of the conflict across Chinese power, identity, and science. Because of the possible political ramifications, Chinese authorities were concerned that a positive DNA linkage between the modern Uyghur people and the ancient mummies would lend the authority of science to the Uyghur contention that they were the direct descendants of the initial colonizers of the Tarim Basin and, therefore, the legitimate heirs to that region.

In 1993, the Chinese government allowed Mair and Paolo Francalacci to take DNA specimens from these ancient mummies. Although the government apparently changed its mind and subsequently seized the specimens before they could be analyzed, enough material was surreptitiously slipped out of the country to permit the extraction of mitochondrial DNA from one individual. Francalacci was able to genetically tie that mummy to a lineage typically encountered in Western Europe.[52] For years the Chinese government successfully blocked additional testing until it funded laboratories at Jilin University in Changchun and at Fudan University in Shanghai to conduct genetic tests on twenty additional Xiaohe mummies. The results, published in 2010 in *BMC Biology*, concluded that genetic evidence showed that the paternal Y-chromosome DNA came predominantly from Eastern Europe, South Asia, and Siberia, and that the maternal mitochondrial DNA could be tied to East and West Eurasian lineages, but rarely East Asia.[53] Radiocarbon dating indicated that the lowest layer of mummies dated to approximately 4000 BP (radiocarbon years).[54] In other words, science has documented that a culturally and genetically independent civilization was already thriving in the Tarim Basin at a time that ethnic Han oral history contends was the beginning of Chinese civilization.

This political discontinuity between science and Chinese cultural history spread to the United States. As described in the *New York Times*, the Beauty of Xiaohe and other Tarim Basin artifacts were, with the agreement of the Chinese government, in the United States in 2011 on tour. With stops at Santa Anna, California's Bowers Museum and Houston's Museum of Natural Science, the exhibition's last location was scheduled for the University of Pennsylvania Museum of Archaeology and Anthropology. However, prior to its opening at the Penn Museum, the Chinese government prohibited the mummies from being shown without providing an explanation. Mair finally received permission from the Chinese government for the mummies to be shown for the first month of the scheduled four-month exhibition.[55] "The unexpected appearance of non-Chinese-seeming cultures and bodies in this region is being treated a bit like the way some American Indian tribes treated the 1996 discovery of Kennewick Man in Washington State," commented Lothar von Falkenhausen, an art historian at the University of California at Los Angeles.[56] The Caucasian-featured Beauty of Xiaohe potentially demonstrated that the ethnic Han Chinese are not historically indigenous to what is now China's Xinjiang province, the home of ethnic Uyghur Muslims who have protested Chinese government and ethnic Han cultural control and are now reportedly massively incarcerated by the government.

# Conclusions

I N GULLIVER'S TRAVELS, JONATHON Swift suggested that the only allow-
able way to crack an egg in Lilliput was from the small end. Those who surreptitiously cracked eggs from the big end against the prevailing cultural and
political standard were known as "big-enders." This human effort to provide a
structure—even a seemingly arbitrary one—for relating to the natural world
transcends any particular culture. Social philosophers have tried to capture the
underlying frames of reference that govern these human interactions—such as
Martin Heidegger and enframing; Michel Foucault and episteme; Sandra Harding and situated knowledge; and Vine Deloria Jr. and American Indian creationism. Despite their many differences, both myth and science share the common
attribute of being cultural frames of reference that address human relationships
with the material world.

For American Indian, feminist, anticolonialist, Marxist, and other historically non-mainstream Western theorists—those who might be considered modern big-enders—the practice of science has been a hegemonic reflection of the
values of a culturally dominant bourgeois/masculine/imperialist class. In their
views, scientific statements reveal as much about political reality as they do the
physical universe. The difficulty, however, faced by these modern big-enders is
that their logic for denying an epistemological advantage to a culturally dominant science also supports not awarding a privileged position to their own
epistemology. To argue that First American science provides only one of many
epistemological vantage points is to argue that there are no privileged vantage
points. This is the challenge faced by both American Indians and First American
science in an intellectually pluralistic American culture. This does not mean
that the modern big-enders are all situated external to the practice of science.
Within mainstream First American science there are epistemological confrontations that are just as fierce as those between science and American Indian traditions. As was discussed earlier, archaeology and anthropology—and to much
lesser extents, linguistics, physics, and genetics—jockey for the pole position
in being the cultural authority for First American science. In the postmodern
world everyone is a big-ender; there is no stable ground on which to place an
epistemological fulcrum.

Much of the conflict between the practice of First American science and the oral traditions of American Indians is conducted in the material world; it is seemingly a confrontation over control of human remains and artifacts. However, at its core, it is an epistemological conflict over how a heritage is to be valued, whether the applied epistemological metric is to be the material quantifications of the Enlightenment or the myths of ancient culture.

The social and scientific politics surrounding the First Americans has made it difficult to separate the issues associated with the original "those people," whoever they were, from the modern American Indian "these people." Consequently, addressing the original First American migration to the New World necessarily means that a First American scientist is, presumptively, also addressing the origins of the modern American Indian. Before concluding that a First American arrival is also a modern American Indian arrival, a scientist would typically require a material basis for such a conclusion. However, because of the unique cultural authority granted by American society, American Indian traditions and oral histories, that is, their cultural myths, have legally been accorded equal epistemological status with the findings of science in the disposition of many ancient human remains and funereal artifacts.

The distinction between myth and science that the Western world has commonly believed to exist is not always apparent in the search for the First Americans. Until the mid-twentieth century, First American science was able to produce material evidence with supporting theories that explained the inherent intellectual superiority of the genetic stock of white Europeans. In this instance, there was a convergence between the findings of modern science and the dominant cultural myth of white superiority. Now that racial science has been discredited, the modern world is left only with a myth of racial inequality. Similarly, Plato's myth of Atlantis endured and was legitimized as the source of the First Americans by prominent nineteenth-century scientists. Now Atlantis has also lost its aura of scientific acceptability.

For those who discount the social utility of myths, this occasional blurring between science and myth weakens the credibility of science as a reliable mechanism for relating to the natural world. This should not, however, be interpreted as endorsing an equivalence between science and myth. As difficult as it may be to dislodge scientific theories such as Clovis-first that no longer are supportable, science does espouse a methodological mechanism for change. Creation myths—the conservation of a particular heritage—has no such accepted process for change. As resistant to change as First American science appears to be, there does reach a tipping point when a theory such as Clovis-first is acknowledged to be wrong.

Although existential myths can certainly evolve over time, they typically have no agreed-upon set of procedures through which they might be disconfirmed.

There is a comforting totality in American Indian myths that is absent in First American science. As laid out by the archaeologist James Adovasio, the puzzle to be addressed is: "who the hell are these people, where did they come from, and when did they get here?"[1] American Indian myths provide complete answers for all questions. One tribe's myth may differ from that of another, but that is acceptable since the myths are typically particular to that specific tribe. Each culture's answers may be different. Science has a larger challenge. It cannot tolerate different answers. Where American Indian myths can be particular, First American science must be universal. Moreover, First American science is preserved via the written report; its methods are preserved in situ at the excavation site. American Indian myth at its heart is preserved orally; there is little of material substance to be preserved. When myths are committed to writing, all literate people can share and interpret them. For cultural authorities who are empowered to make and interpret truth statements, a source of power can then be challenged.

American Indian myths and First American science have such substantive differences that it is difficult to imagine a circumstance under which there could be an epistemological reconciliation. For American Indians, their existential myths are statements of truth. Each tribe can have its own myth that provides it an ontologically stable answer to the Adovasio questions. For First American science, however, the answers may (and do) change, but there is only one truth permitted at a time; scientists are serial ontologists. It is the scientific methodology that is enshrined as the stable element. For American Indians, it is the answer that is to be culturally protected; for First American scientists, it is the path that is critical. This seeming clarity in differences is, unfortunately, frequently blurred when one domain attempts to co-opt the authority of the other. When nineteenth-and early twentieth-century scientific methodologies were used to support the racist doctrines of the dominant culture, the hierarchy of races was a cultural truth; the then-methods of science were used to validate what was believed already known. The history of science would indicate that it may only be in distant retrospect that it is possible to distinguish between myth and science.

A second area of apparently irreconcilable difference between American Indian myths and First American science is that they are not actually addressing the same questions. Adovasio's first question—"who the hell are these people"— appears straightforward. The answer requires situating "these people" between their ancestral home and their modern descendants, if any. Yet for American Indians, the answer is typically couched in terms of cultural relationships; for First

American scientists, the answer is principally one of biology. Proving a cultural continuity does not address a biological relationship, and vice versa. That the two domains are talking past each other was readily apparent in the confrontations over control of the remains of the Kennewick Man. For First American scientists, there was initially sufficient anatomical variation between the Kennewick Man and modern American Indians to question a relationship. For American Indians, their myths confirmed that they, and only they, had inhabited the area where the remains were found, so that there was an established cultural continuity; there could have been no European ancestral interloper.

There is an inherent culturally based distinction between American Indians and modern, scientifically based society. American Indian myths are a part of their social fabric and are not discoverable. They have a status that is believed to be independent of any human process for acquiring knowledge. There is a social value to their creation myths that structure and stabilize their society. Something accorded the status of a creation myth is not irrational but is considered a fundamental part of reality. For such cultures, myth creation is an ontologically definitive event. Myths are nonargumentative. What initially may have been figurative is now imbued with literality. American Indian myths and First American science each represent a particular approach to existence. It is doubtful that the cognitive values important to either could be shown to be commensurate with the other.

In the search for the First Americans, American Indian tribes are concerned with protection of their existential myths; they wish to conserve a particular heritage. Although scientists portray their discipline as one that is open to change, the actual practice of First American science has been one that is significantly resistant to change. There is a great body of science studies literature that describes scientific change as occurring in paradigm shifts, epistemological ruptures, punctuated equilibria, and discontinuities.[2] It is difficult to envision revolutionary creation myth changes except, perhaps, when externally imposed by an alien cultural authority. For American Indians, First American science is just such an unwelcome external mechanism that attempts to destabilize their existential myths.

It has been suggested that science is a form of metaphysical and methodological commitment.[3] There is also a metaphysical and methodological commitment associated with any particular myth. Michel Foucault coined the term epistemological grid to define the underlying, and often unacknowledged, value structure against which these cultural commitments are made. Neither First American science nor American Indian myths can be considered to be situated independently of such a pervasive grid; the challenge in the search for the First Americans is that these epistemological grids are not identical and are most likely incommensurable. Furthermore, both science and myth are frequently uncooperative in that they do not stay ontologically stable. As the social context changes, the values embedded in the epistemological grid change as well— whether for science or myth.

Methods constrain truth and are culturally and historically contingent. Verisimilitude—Foucault's regimes of truth—is a human endeavor and is subject to human foibles. That does not, however, translate "wrong" into a constructivist's belief that all social solutions are equally acceptable. First American science, more so than American Indian myth, is still bound by the modern perception of the physicality of nature. In American society, numbers are the ultimate scientific method, the universal approach for unraveling the mysteries of nature. It is difficult to imagine a method for mathematizing a myth.

Scientists organize themselves separately from the engineering disciplines, and further divide themselves into separate domains such as physics, biology, and chemistry. This structure is evident in the departmental divisions in academia. These socially constructed groupings of concepts and human skills, however, do not reflect how science is actually practiced. Science studies has not only documented that the practice of modern science cannot be conducted without crucial contributions from the technologies of the engineering disciplines, but also shows that the scientific approaches to understanding the natural world consist not just of physics or archaeology, but rather suites of sciences addressing specific puzzles and research programs. Despite the declarations by archaeologists that the search for the First Americans is inherently an archaeological issue, solving the conundrum of the First Americans requires not only the application of subsets of archaeology, anthropology, linguistics, bioanthropology, and physics, but also requires an epistemological grid—the context of a specific set of dominant cultural values—that will accept the results of science. American Indians would certainly not agree that the search for the First Americans is solely an archaeological issue, nor even solely a science issue. In the Native American Graves Protection and Repatriation Act (NAGPRA), the US Congress has, in fact, dictated in law that the search for the First Americans is not to be solely a matter for First American science.

In spite of the intellectual force brought to bear from the application of an entire ensemble of sciences to the First American puzzle, for seventy years the now-discredited Clovis-first theory was repeatedly validated. In retrospect, First American scientists appear to have been more focused on continuously confirming the Clovis-first model than searching for material artifacts and constructing testable alternative theories. Clovis-first became so dominate (or more likely, the fear of challenging the elite First American scientists that supported the theory was so great) that it became a part of the *episteme* of the First American search. Helen Longino concluded that the results of a scientific investigation "can be taken as evidence for hypotheses only in the context of some set of background beliefs," a part of the epistemological grid.[4] So objectified did the Clovis-first model become, so blurred was the distinction between theory and material evidence, that it assumed the authority of an artifact itself.

The questionable nature of scientific conclusions has become an increasingly popular research topic. In his thirteen-year study of medical research, "Why Most Published Research Findings Are False," John Ioannidis concluded that "for most study designs and settings, it is more likely for a research claim to be false than true. Moreover, for many current scientific fields, claimed research findings may often be simply accurate measures of the prevailing bias."[5] In *Wrong*, David Freeman explained that researchers at Harvard and the National Bureau of Economic Research "examined papers from a range of economics journals and determined that approximately none of them had conclusively proved anything one way or the other."[6] Moreover, the proof of many scientific findings is inextricably intertwined with the acceptance of the methods through which the findings are produced. In some First American sciences such as archaeology, the *only* basis for validating the findings may well be legitimization of the methods employed, not validation of the findings. Echoing Polanyi's theory of tacit knowledge, Richard Feynman concluded that "we have a habit in writing articles published in scientific journals to make the work as finished as possible, to cover up all the tracks, to not worry about the blind alleys or describe how you had the wrong idea first, and so on."[7] The economic theory of suboptimization describes how a person strives to maximize an individual gain that may ultimately be to the detriment of a larger and more important interest. How First American scientists practice their discipline demonstrates a form of suboptimization. An overzealous protection of one particular theory may well be to an individual's benefit; however, First American science may well be the ultimate loser through an erosion of confidence in science as a whole.

There is a continuity in how observations occur in science that ranges across the material spectrum of static display, controlled demonstration, and theoretical experimentation. A commonly held belief is that the practice of science must involve some form of experimentation. In First American science there is certainly a form of observation, but it is difficult to identify a practice that might be considered experimentation. In the domains of bioanthropology, physics, genetics, and linguistics that support First American science, there would appear to be no experimentation. Craniometry, carbon dating, DNA typing, and language analysis are sufficiently routine to be considered technological measurement demonstrations rather than scientific experiments. For archaeology and anthropology there is little distinction that can be made between what constitutes a First American science laboratory and what is a museum. An excavation site has the characteristics of both display and demonstration, but it is doubtful that it can be considered experimentation.

Further compounding what constitutes an observation in First American science is that exactly where an observation may be made is critically dependent on who owns the land. Many of the potentially best excavation sites are located on land controlled by American Indian tribes or federal agencies that frequently accommodate American Indian interests. As a result, the location of a First American laboratory is not a value-neutral process. The context through which the First American search occurs is the result of a social negotiation between First American scientists and the very groups that frequently oppose them.

The application of molecular analysis in physics has greatly strengthened the capability of First American science to provide dates for its material artifacts. With physics widely regarded as the archetype of the hard sciences, this physics-associated dating capability is also associated with an increase in public acceptance. Yet, there are substantive limitations to the contribution that physics has made. Typically, the popular press reports the results of First American science carbon dating as a relatively precise number. It is doubtful, however, that this same public understands that the carbon dating of an ancient human bone, for example, is dependent, initially, on what is most likely: (1) a very accurate calculation of carbon 14 decay in the bone but that, unfortunately, (2) can only be converted to a chronological age by estimating that ancient human's dietary consumption of plants, which in turn (3) is dependent on an estimate of the food-plant's air-temperature-dependent rate of consumption of carbon 14 from the atmosphere, which in turn is (4) dependent on an estimate of the rate of creation of carbon 14 through ionization of the Earth's upper atmosphere by the variable electro-magnetic radiation from the sun, which in turn is (5) calibrated using estimates from the growth cycles of tree rings and from the chemical composition of core samples from the ice caps in Greenland and the Antarctic. This does not mean that carbon dating is not of value; it has provided an invaluable mechanism for improving the chronological estimates associated with the search for the First Americans. It does, however, mean that there are caveats, frequently unarticulated, that should be remembered when considering First American chronologies.

First American science and American Indian myths are each, in its own way, an example of cultural hegemony. However, in the context of modern American culture, a belief in the preeminent value of science has the luxury of being firmly embedded in the general American epistemological grid, while an American Indian myth has social value only in the context of a particular tribe. If a truth statement of either First American science or an America Indian myth was always confirmed, then there would be little controversy; an ontological winner could be declared. Obviously, however, this is not the case. It is difficult to provide scientifically acceptable legitimization for a myth, but science has also demonstrated its own set of challenges. It is these interstitial ontological voids between First American science and American Indian myths that provide an opening for the exercise of political power.

One need not favor American Indian traditions over the findings of First American science to recognize that there is a subtle, but pervasive, cultural grid in the United States against which all values are calibrated. For example, during my vacation stay at the Mountain Creek Lodge at Pipestem State Park in West Virginia, there was a display of American Indian jewelry from prior to the arrival of Europeans that described the makers of the jewelry as "early visitors" to the area.[8] If the dominant American culture can be successful in denying ownership by the American Indian of the material world that they inhabited, whether land or jewelry, then they can also be denied ownership of the ontological.

Politics is a process for resolving unstabilized power relationships, and the United States is not alone in its expression of political power through its search for first inhabitants. For decades, the Chinese government, dominated by the Han ethnic majority, deliberately blocked an open examination of the origins of the four-thousand-year-old European-appearing mummies in the Tarim Basin in China's Xinjiang Uyghur Province because such an investigation has the potential to prove that at least a portion of what is now modern China was initially populated by a people who were both genetically and culturally not east Asian. There is more at stake for the Chinese government than simply losing the epistemological high ground. The Uyghurs—who are predominately Muslim and claim to be the descendants of the Tarim culture—express a desire for independence from Communist and Han China. Denial of an open examination of the mummies is a reflection of power relationships. The politics of power in China intercede in the scientific examination of ancestral roots. In China, the power to control the epistemological debate has been stabilized in the hands of an authoritarian government.

In the democratic United States, power has been directed through the political process in the search for the First Americans more ambiguously than in the Chinese instance. Science has historically played a significant advisory role to the US government; it not only helps to define the problems, but it also provides answers. However, unlike in China, the pluralistic manner in which power is expressed in the United States has resulted in enactment of the Native American Graves Protection and Repatriation Act (NAGPRA) that, in law, grants epistemological equality between American Indian traditions and science in determining the disposition of many ancient human remains and funerary objects. As shown in the United States and Chinese examples, governmental involvement is one of the most blatant external interventions that can be imposed on the scientific process in the search for the First Americans (and Chinese). Governmental power can, and does, dictate the cultural rules that define acceptable epistemological approaches. In enacting NAGPRA, the Congress has decided who gets to assert truth. Political power not only speaks to truth, but truth can be forced to obey.

The politics of power has resulted in First American science being accused of having historically supported an agenda of racism, nationalism, imperialism, and colonialism. With at least some of the historical charges having been substantiated, it is difficult to determine what a proper role might be for First American science in defining and maintaining American existential myths and self-identity. If some subjects are to be off-limits for examination, then it is not clear what First American science has become. Leaders of the American Anthropological Association unsuccessfully attempted to delete science from their mission statement in order to permit non-science-based epistemological approaches and to preclude the continuing accusations of being a tool of an imperialist American culture. This would appear to be an admission by elite anthropologists of AAA that at least the anthropological element of science, as an independent entity, could no longer be trusted to freely examine First American myths and identities with the methodologies of science.

Who owns the past is a function of who owns the power for defining the process for determining the past. Creation of a past is of value as a mechanism for controlling the future. With the legislative provisions included in NAGPRA and with the creation of, and substantial funding provided by, the American Indian heritage preservation offices, there has been a significant shift in the balance of political power between First American science and American

Indian traditions. Academic First American scientists now appear at odds with their business counterparts who are responsive to the heritage offices that form their customer base and who are less enamored with the traditional peer review process revered by the academic community. The erosion of the power of First American science is also reflected in the occasional appearance of tribal affiliations of authors as a credential worthy of note in academic publications. At stake is nothing less than control over interpretation of the American past.

The substantial distrust between many First American scientists and American Indians has manifested itself in a very public manner. The confrontation over the remains of the Kennewick Man was worthy of a Greek passion play. The ancient human remains were discovered by two inebriated men during an attempt to avoid paying an admission fee, and subsequently included elements of racism, federal agency bias both in favor of and against American Indians, misplaced artifacts, a contest between biology and culture, and an ultimate determination by a federal court.

Archaeology is the discipline most central to the success of First American science, but simultaneously also its weakest link. First, it is widely acknowledged, even among archaeologists, that there is little independent archaeological theory; the discipline must borrow from anthropology for its doctrine. Second, the premier professional association of anthropology, which supplies the intellectual basis for archaeological investigations, considered declaring that the profession is no longer a science. Third, prominent archaeologists have charged that graduates of archaeological programs are sufficiently deficient in science training that they cannot communicate adequately with the physicists and geneticists that are critical to supporting their discipline. Fourth, science has aspirations to more than merely describe the physical world; it endeavors to explain. A central part of that effort is experimentation and repeatability of results. What in archaeology constitutes an experiment? What can be replicated? Fifth, many of the most suitable sites for archaeological excavation cannot be accessed without permission either from American Indian tribes themselves or from federal agencies that have historically shown a propensity to support the political agendas of American Indian tribes. Sixth, funding available for archaeological searches related to the First Americans increasingly has come from the Cultural Resource Management offices of American Indian tribes and these same federal agencies. Seventh, even acknowledging that scientists do not always welcome challenges to their ideas, the archaeology associated with the First American search of the past 150 years appears to have been particularly susceptible to a theoretical rigidity enforced by a small group of prominent scholars variously referred to as "police," or the "bête noire" of new theories. Finally, the archaeology profession associated with the search for the First Americans has fractured. On the one side are the university-based scholars who continue to espouse the traditional science mantras of peer review, objectivity, and a tenure system designed to separate remuneration from scholarly results. On the other side is the ever-increasing number of business-oriented archaeologists that operate in a financially competitive environment, must accommodate a customer base composed in substantial part by American Indian organizations, and whose intellectual products are not peer reviewed.

The practice of science in the search for the First Americans is a complex and flawed process. It must survive in an environment of shifting centers of political power, some of which are inherently hostile to the scientific enterprise. A case can be made that some of the critical elements of First American science, especially archaeology, may no longer even be a science. What is clear, however, after five hundred years of interest by the dominant Euro-American culture—the last 150 years of which have employed the tools of modern science—is that America has made very little progress identifying its first inhabitants. The speculations of an Asian source for the First Americans by José de Acosta in 1590 or by Thomas Jefferson in 1784 may be couched in different terms than those used today, but First American science is not yet able to either confirm or refute their centuries-old opinions. Whether one believes American Indian creation myths or not, it is obvious that there was a First American. What remains unsettled is not only who these people were and when they came, but also the very science associated with the search for the First Americans.

## Introduction

1. McGuire, *Archaeology as Political Action*, 141.
2. Obama, *Inaugural Address*.
3. Oreskes, *The Rejection of Continental Drift*; Biagioli, *Galileo, Courtier*.
4. Feyerabend, *Against Method*, 170.
5. Wilmsen, "An Outline of Early Man Studies in the United States," 13.
6. Meltzer, "Why Don't We Know When the First People Came to North America?," 26.
7. See, for example, the work of Stephen Jay Gould addressing Non-Overlapping Magisteria (NOMA) in *Leonardo's Mountain of Clams and the Diet of Worms*, contending that religion and science were not inherently in conflict.
8. Curtin, *Creation Myths of Primitive America*.
9. Foucault, *The Order of Things*.
10. Longino, *Science as Social Knowledge*, 74.
11. Owsley, "From Jamestown to Kennewick," 128.
12. Deloria, *Red Earth, White Lies*.
13. Adovasio, *The First Americans*, 124.

## Chapter 1

1. Von Franz, *Creation Myths*, 1.
2. Deloria, *Red Earth, White Lies*, 37.
3. The relationship between the Judeo-Christian ethos and the Western concern with dominion over nature is a recurring theme in science studies literature, but many attribute an early articulation to Lynn White, "The Historical Roots of Our Ecological Crisis," 1203–07.
4. Villa, *Arendt and Heidegger*, 190.
5. Deloria, *Red Earth, White Lies*, 39.
6. Von Franz, *Creation Myths*, 132–33.
7. Pijoan, *American Indian Creation Myths*, 8–9.
8. Deloria, *Red Earth, White Lies*, 233.
9. Pijoan, *American Indian Creation Myths*, 8.
10. Pijoan, *American Indian Creation Myths*, 8–9.

11. Deloria, *Red Earth, White Lies*, 11.

12. Curtin, *Creation Myths of Primitive America*, 487.

13. Curtin, *Creation Myths of Primitive America*, xiii.

14. Curtin, *Creation Myths of Primitive America*, xx.

15. Pijoan, *American Indian Creation Myths*. Pijoan provides a very readable catalogue of American Indian myths.

16. Von Franz, *Creation Myths*, 28. This myth is reported by Knud Rasmussen in his book *The Eagle's Gift* and discussed by von Franz.

17. Curtin, *Creation Myths of Primitive America*, xxii.

18. Curtin, *Creation Myths of Primitive America*, xxv.

19. Hawking, *A Brief History of Time*, 1.

20. Latour and Woolgar, *Laboratory Life*.

21. Naddaf, "Allegory and the Origins of Philosophy," 102.

22. Naddaf, "Allegory and the Origins of Philosophy," 121, footnote 10.

23. Morgan, *A Myth and Philosophy from the Presocratics to Plato*, 1. See also discussions in Lloyd, *Demystifying Mentalities*, and Hatab, *Myth and Philosophy*.

24. Water Beetle exhibit observed by the author at the Smithsonian Institution National Museum of the American Indian, July 29, 2010.

25. Earth Medicine exhibit observed by the author at the Smithsonian Institution National Museum of the American Indian, July 29, 2010.

26. Exhibit observed by the author at the Smithsonian Institution National Museum of the American Indian, July 29, 2010.

27. Exhibit observed by the author at the Smithsonian Institution National Museum of the American Indian, July 29, 2010.

28. Contact exhibit observed by the author at the Smithsonian Institution National Museum of the American Indian, July 29, 2010.

# Chapter 2

1. Adovasio, *The First Americans*, 4.

2. For the purposes of this discussion, I have chosen to use the term Euro-American to describe the culture in the geographical area that has become the United States. While there are certainly distinctions that might be made between European and American views in the search for the First Americans, they are of minor significance when compared with those held by American Indians as discussed in the previous chapter. There is a continuity of culture in what is captured here as Euro-American, as opposed to the significant discontinuity presented by American Indian beliefs.

3. Goebel, Waters, and O'Rourke, "The Late Pleistocene Dispersal of Modern Humans in the Americas," 1498.

4. Wilmsen, "An Outline of Early Man Studies in the United States," 3.

5. Wilmsen, "An Outline of Early Man Studies in the United States," 3.

6. For a pictorial reference, see maps from Dillehay, "Probing Deeper into First American Studies," 973.

7. See d'Anghiera, *De Orbe Novo*.

8. Meltzer, *First Peoples in a New World*, 64.

9. Winsor, "The Progress of Opinion Respecting the Origin and Antiquity of Man in America," 369; cited in Wilmsen, "An Outline of Early Man Studies in the United States," 3.

10. Browman and Williams, *New Perspectives on the Origins of Americanist Archaeology*, 35. See de Acosta, *Historia natural y moral de las Indias*.

11. Adovasio, *The First Americans*, 5.

12. Paracelsus quoted in *Memoirs Read Before the Anthropological Society of London*, 354.

13. Wilmsen, "An Outline of Early Man Studies in the United States," 3–4.

14. Wilmsen, "An Outline of Early Man Studies in the United States," 4. Cotton Mather quotation is Wilmsen citing Samuel G. Drake.

15. Hobbes, "Of the Natural Condition of Mankind as Concerning Their Felicity and Misery."

16. Leclerc, *Histoire naturelle, générale et particulière*.

17. Jefferson, *Notes on the State of Virginia*.

18. Pfefferkorn, quoted in Wilmsen, "An Outline of Early Man Studies in the United States," 3–4.

19. Wilmsen, "An Outline of Early Man Studies in the United States," 7.

20. Adovasio, *The First Americans*, 35.

21. Adovasio, *The First Americans*, 10.

22. A modern reprint is available through Applewood Books, Carlisle, Massachusetts.

23. Adovasio, *The First Americans*, 16–17.

24. Wilmsen, "An Outline of Early Man Studies in the United States," 6.

25. Donnelly, *Atlantis*, 198.

26. Donnelly, *Atlantis*, 209.

27. Adovasio, *The First Americans*, 18.

28. Foster, *Pre-Historic Races of the United States of America*.

29. Winsor, "The Progress of Opinion Respecting the Origin and Antiquity of Man in America," 369.

30. This is thirteen thousand years ago in chronological years from the present. As later described in more detail, this is approximately calibrated to 11,500 radiocarbon years before the present, with the "present" defined as 1950.

31. Meltzer, *First Peoples in a New World*, 184.

32. Dillehay, "Probing Deeper into First American Studies," 971.

33. Holmes, "Handbook of Aboriginal American Antiquities," 75.

34. Hrdlička, "The Origin and Antiquity of the American Indian," 481–84.

35. Wilmsen, "An Outline of Early Man Studies in the United States," 10.

36. Adovasio, *The First Americans*, 97.

37. Meltzer, "Why Don't We Know When the First People Came to North America?," 25. "B.P." stands for "Before Present" and indicates that the years are not chronological years, but rather are radiocarbon years, which do not exactly equate to chronological years. Moreover, Before Present does not mean before 2021, but rather before 1950 when radiocarbon dating began being used. This is discussed in more detail later. 12,000 BP translates approximately to 13,500 chronological years ago.

38. Adovasio, *The First Americans*, xviii.

39. Meltzer, "Why Don't We Know," 25.

40. Gibson, "EXCERPTS: Charlie Gibson Interviews Sarah Palin."

41. United States National Oceanic and Atmospheric Administration (NOAA), www.ncdc.noaa.gov/ (accessed June 26, 2019).

42. Fiedel, *Prehistory of the Americas*.

43. Adovasio, *The First Americans*, 46.

44. Deloria, "Foreword," xvii.

45. Deloria, *Red Earth, White Lies*, 45.

46. Deloria, *Red Earth, White Lies*, 97.

47. Weaver, "Indian Presence with No Indians Present," 25.

48. Weaver, "Indian Presence with No Indians Present," 25.

49. Arnold, "The Ice-Free Corridor," iii.

50. See Mandryk, "Paleoecology as Contextual Archaeology," and Stanford and Bradley, *Across Atlantic Ice*.

51. Erlandson, Rick, Vellanoweth, and Kennett, "Maritime Subsistence at a 9300-Year-Old Shell Midden on Santa Rosa Island, California," 255–65.

52. Anderson and Gillam, "Paleoindian Colonization of the Americas," 50.

53. Meltzer, "Why Don't We Know," 35.

54. Meltzer, "Why Don't We Know," 35.

55. Anderson and Gillam, "Paleoindian Colonization of the Americas," 49.

56. Anderson and Gillam, "Paleoindian Colonization of the Americas," 62.

57. Anderson and Gillam, "Paleoindian Colonization of the Americas," 63–65.

58. Bradley and Stanford, "The North Atlantic Ice-Edge Corridor," 459–78.

59. Dillehay, "Probing Deeper into First American Studies," 973.

60. Chatters, *Ancient Encounters*, 260.

61. Chatters, *Ancient Encounters*, 261.

# Chapter 3

1. Curry County, New Mexico, Chamber of Commerce, "Clovis/Curry County History," www.clovisnm.org/pages/history (accessed June 26, 2019).

2. Katz, *The History of Blackwater Draw*, 17.

3. Katz, *The History of Blackwater Draw*, 16.

4. Katz, *The History of Blackwater Draw*, 8.

5. Katz, *The History of Blackwater Draw*, 9.

6. This is Dr. John Carter being quoted from 1997 recollections of his participation in the 1936–37 participation in Blackwater Draw excavations, cited in Katz, *The History of Blackwater Draw*, 8.

7. Marshall, "Clovis Counterrevolution," 738–41.

8. Marshall, "Clovis Counterrevolution," 738.

9. Karl Butzer, quoted in Marshall, "Clovis Counterrevolution," 739.

10. Wilford, "'American' Arrowhead Found in Siberia."

11. See Sauer, "A Geographic Sketch of Early Man in America," 529–73; Martin, "Pleistocene Overkill," 32–38.

12. Adovasio, *The First Americans*, 123–28.

13. Zorich, "Undiscovery of the Year," 30.

14. Marshall, "Clovis Counterrevolution," 741.

15. Marshall, "Clovis Counterrevolution," 741.

16. Katz, *The History of Blackwater Draw*, 25; "Russian Money Feeds International Art Boom."

17. Katz, *The History of Blackwater Draw*, 25.

18. David Kilby, Eastern New Mexico University home page, http://academic.enmu.edu/kilbyd/fieldprojects.html (accessed May 6, 2011; he is now at Texas State University); "Russie.net, le web franco-russe," http://www.russie.net/L-exposition-de-Modigliani-au (assessed June 27, 2019).

19. David Kilby, quotation from his Eastern New Mexico University home page, http://academic.enmu.edu/kilbyd/fieldprojects.html (accessed May 6, 2011); he is now at Texas State University.

20. *Clovis News Journal*, http://www.cnjonline.com/articles/dig-34373-site-isequilla.html (accessed May 6, 2011).

21. Meltzer, "Why Don't We Know When the First People Came to North America?," 26.

22. Huckell and Kilby, *Readings in Late Pleistocene North America and Early Paleoindians*, vii.

23. Jelinek, "Perspectives from the Old World on the Habitation of the New," 45.

24. Anderson and Gillam, "Paleoindian Colonization of the Americas," 61–62.

25. Jelinek, "Perspectives from the Old World on the Habitation of the New," 47.

## Chapter 4

1. Popper, *The Logic of Scientific Discovery*.

2. Meltzer, "Why Don't We Know When the First People Came to North America?," 31.

3. Meltzer, Grayson, Ardila, Barker, Dincauze, Haynes, Mena, Núñez, and Stanford, "On the Pleistocene Antiquity of Monte Verde, Southern Chile," 295.

4. Meltzer, Grayson, Ardila, Barker, Dincauze, Haynes, Mena, Núñez, and Stanford, "On the Pleistocene Antiquity of Monte Verde, Southern Chile," 295.

5. Meltzer, Grayson, Ardila, Barker, Dincauze, Haynes, Mena, Núñez, and Stanford, "On the Pleistocene Antiquity of Monte Verde, Southern Chile," 298.

6. Meltzer, Grayson, Ardila, Barker, Dincauze, Haynes, Mena, Núñez, and Stanford, "On the Pleistocene Antiquity of Monte Verde, Southern Chile," 296–98.

7. Dillehay, *Monte Verde, A Late Pleistocene Settlement in Chile*.

8. "Monte Verde Under Fire," *Archaeology*.

9. The full author citation is provided here since the authors constituted the group of senior scientists who performed the site inspection: David J. Meltzer (Southern Methodist University), Donald K. Grayson (University of Washington), Gerardo Ardila (Universidad Nacional de Colombia), Alex W. Barker (Dallas Museum of Natural History), Dena F. Dincauze (University of Massachusetts), C. Vance Haynes (University of Arizona), Francisco Mena (Museo Chileno de Arte Precolombino), Lautaro Núñez (Universidad Catolica del Norte), and Dennis J. Stanford (Smithsonian Institution), "On the Pleistocene Antiquity of Monte Verde, Southern Chile." Tom Dillehay and James Adovasio also attended the site visit but did not join in coauthoring the article because they were involved in the site excavation. Robson Bonnichsen attended the preliminary meeting in the United States but did not travel to the site.

10. Weaver, "Indian Presence with No Indians Present," 23.

11. Fiedel, "Artifact Provenance at Monte Verde."

12. Tom D. Dillehay (University of Kentucky), Michael B. Collins (University of Texas), Mario Pino (Universidad Austral de Chile), Jack Rossen (Ithaca College), Jim Adovasio (Mercyhurst College), Carlos Ocampo (Universidad de Chile), Ximena Navarro (Universidad Catolica de Chile), Pilar Rivas (UCA, S.A.), David Pollack (Kentucky Heritage Council), A. Gwynn Henderson (Kentucky Archaeological Survey), Jose Saavedra (Casa de la Cultura, Chile), Patricio Sanzana (Consejo Nacional de Desarollo Indigena, Chile), Pat Shipman (Penn State University), Marvin Kay (University of Arkansas), Gaston Munoz, Anastasios Karathanasis (University of Kentucky), Donald Ugent (Southern Illinois University), Michael Cibull (University of Kentucky), and Richard Geissler (University of Kentucky), "On Monte Verde: Fiedel's Confusions and Misrepresentations," 1, www.uky.edu/Projects/MonteVerde/monteverde.pdf (accessed June 26, 2011).

13. Dillehay, "The Battle of Monte Verde," 32, cited in Weaver, "Indian Presence with No Indians Present," 23–24.

14. Meltzer, "Why Don't We Know When the First People Came to North America?," 38.

15. Adovasio, *The First Americans*, xiii–xiv.

16. Adovasio, Gunn, Donahue, and Stuckenrath, "Meadowcroft Rockshelter, 1977: An Overview," 261.

17. Adovasio, *The First Americans*, xv.

18. Adovasio, *The First Americans*, xv.

19. Observed by the author during a site visit on July 17, 2010.

20. Heinz History Center, www.heinzhistorycenter.org/about (accessed June 27, 2019).

# Chapter 5

1. Bleiler in the introduction to Donnelly, *Atlantis*, v.
2. Donnelly, *Atlantis*, 22.
3. Adovasio, *The First Americans*, 27–28.
4. Donnelly, *Atlantis*, xviii–xix.
5. Donnelly, *Atlantis*, xii.
6. Donnelly, *Atlantis*, xiv.
7. Donnelly, *Atlantis*, 98.
8. Donnelly, *Atlantis*, 348.
9. Donnelly, *Atlantis*, xviii–xix.
10. Donnelly, *Atlantis*, xix.
11. Donnelly, *Atlantis*, 47.
12. Donnelly, *Atlantis*, Mycenae sketch, 354; Central American sketch, 355.
13. Donnelly, *Atlantis*, 274.
14. Adovasio, *The First Americans*, 27–28.
15. Donnelly, *Atlantis*, 116.
16. Donnelly, *Atlantis*, 219.
17. Donnelly, *Atlantis*, 435–36.
18. Foster, *Pre-Historic Races of the United States of America*, 397–98.
19. Foster, *Pre-Historic Races of the United States of America*, 398. Foster does not give Catlin's first name, but it must certainly be George Catlin, the historian and painter of American Indian culture. Foster cites the work as being from "Revue des Deux Mondes," March 1867, 154.
20. Meltzer, "Why Don't We Know When the First People Came to North America?," 29–30.
21. Strassberg, *A Chinese Bestiary*, xiii.
22. Strassberg, *A Chinese Bestiary*, 10.
23. Strassberg, *A Chinese Bestiary*, 55.
24. Strassberg, *A Chinese Bestiary*, 55.
25. Strassberg, *A Chinese Bestiary*, 35.

# Chapter 6

1. International Union for Quaternary Research (IUQR) letter of 2006, cited in Meltzer, *First Peoples in a New World*, 29.
2. Meltzer, *First Peoples in a New World*, 29.
3. Adovasio, *The First Americans*, 95.
4. Meltzer, "Why Don't We Know When the First People Came to North America?," 37.
5. Meltzer, "Why Don't We Know When the First People Came to North America?," 25.
6. McGuire, *Archaeology as Political Action*, xii.

7. Chinard, "Jefferson and the American Philosophical Society," 263–76.

8. Adovasio, *The First Americans*, 95.

9. This is attributed to Feynman, but as with many aphorisms attributed to famous sources, there is no definitive documentation as to time and place of utterance.

10. American Anthropological Association website, http://www.americannanthro.org/ConnectWithAAA/Content.sapx?ItemNumber=1665&navItemNumber=586 (accessed July 13, 2019).

11. Attachment to November 23, 2010, email by Peter Neal Peregrine, http://www.unl.edu/rhames/AAA/AAA-LRP.pdf (accessed December 13, 2010).

12. Clark, "Should One Size Fit All?," 39, 40, 41.

13. Wade, "Anthropology as Science."

14. Virginia Dominguez, cited in Wade, "Anthropology as Science."

15. Dreger, "No Science, Please, We're Anthropologists."

16. Stuart Plattner, cited in Dreger, "No Science, Please, We're Anthropologists."

17. Raymond Hames, cited in Dreger, "No Science, Please, We're Anthropologists."

18. Raymond Hames, cited in Berrett, "Anthropology Without Science."

19. Berrett, "Anthropology Without Science."

20. Longino, *Science as Social Knowledge*, 98.

21. Said, *Culture and Imperialism*, 152, cited in Weaver, "Indian Presence with No Indians Present," 28, note 4.

22. Weaver, "Indian Presence with No Indians Presents," 15.

23. AAA Statement of Purpose, https://www.americananthro.org/ConnectWith-AAA/Content.aspx?ItemNumber=1650&navItemNumber=760 (accessed August 10, 2019).

24. Meltzer, *First Peoples in a New World*, 202.

25. Dillehay, "Probing Deeper into First American Studies," 971.

26. Adovasio, The First Americans, xvi.

27. Meltzer, "Why Don't We Know When the First People Came to North America?," 29.

28. Deloria, "Foreword," xviii.

29. Meltzer, *First Peoples in a New World*, 12.

30. Kintigh, "Repatriation as a Force of Change in Southwestern Archaeology," 204.

31. Deloria, *Red Earth, White Lies*, 49.

32. Killion, *Opening Archaeology*, 9.

33. Clark, "Should One Size Fit All?," 41.

34. Killick and Goldberg, "A Quiet Crisis in American Archaeology," 8.

35. Killick and Goldberg, "A Quiet Crisis in American Archaeology," 9.

36. Clark, "Should One Size Fit All?," 39.

37. Shott, "Crises and Solutions in American Archaeology," 38.

38. Killick and Goldberg, "A Quiet Crisis in American Archaeology," 10.

39. Clark, "Should One Size Fit All?," 41.

40. Shott, "Crises and Solutions in American Archaeology," 38.

41. Adovasio, *The First Americans*, xvi.

## Chapter 7

1. Chatters, *Ancient Encounters*, 112

2. Goebel, Waters, and O'Rourke, "The Late Pleistocene Dispersal of Modern Humans in the Americas," 1498.

3. Fiedel, *Prehistory of the Americas*, 39.

4. See Chatters, *Ancient Encounters*, 192, for a map of the North American locations citing the specific remains associated with each.

5. Dongoske and Zimmerman, "Working Together on Race and Racialism in American Archaeology," 3.

6. Kakaliouras, "Race Is . . . Only as Race Does," 16.

7. Owsley, "From Jamestown to Kennewick," 129.

8. Franz Boas, quoted in Fiedel, *Prehistory of the Americas*, 41.

9. Fiedel, *Prehistory of the Americas*, 351.

10. Fiedel, *Prehistory of the Americas*, 350.

11. Foster, *Pre-Historic Races of the United States of America*, 275.

12. Foster, *Pre-Historic Races of the United States of America*, 275.

13. Foster, *Pre-Historic Races of the United States of America*, 275.

14. Foster, *Pre-Historic Races of the United States of America*, 279.

15. Foster, *Pre-Historic Races of the United States of America*, 277.

16. Foster, *Pre-Historic Races of the United States of America*, 280.

17. Adovasio, *The First Americans*, 19.

18. Gould, *The Mismeasure of Man*.

19. Lewis, DeGusta, Meyer, Monge, Mann, and Holloway, "The Mismeasure of Science," 1.

20. Lewis, DeGusta, Meyer, Monge, Mann, and Holloway, "The Mismeasure of Science," 1.

21. Kakaliouras, "Race Is . . . Only as Race Does," 17.

22. Owsley, "From Jamestown to Kennewick," 133

23. Chatters, *Ancient Encounters*, 126.

24. Chatters, *Ancient Encounters*, 175.

25. FORDISC 3.1, fac.utk.edu/fordisc-3-1-personal-computer-forensic-discriminant-functions/ (accessed July 3, 2019).

26. Opfer, "Body of Evidence," *Popular Science*, July 2010, 34.

27. Kakaliouras, "Toward a 'New and Different' Osteology," 119.

28. Kakaliouras, "Toward a 'New and Different' Osteology," 115.

29. Kakaliouras, "Toward a 'New and Different' Osteology," 126.

30. Fiedel, *Prehistory of the Americas*, 25.

31. Chatters, *Ancient Encounters*, 219–26.

32. Dentochronology (the study of teeth) is to be distinguished from dendrochronology (the study of tree rings), both important in the search for the First Americans.

33. Foster, *Pre-Historic Races of the United States of America*, 38.

34. Meltzer, "Why Don't We Know When the First People Came to North America?," 27.

35. Greenberg, Turner, and Zegura, "The Settlement of the Americas," 477.

36. Meltzer, "Why Don't We Know When the First People Came to North America?," 27.

37. Campbell, *American Indian Languages*, 101.

38. Campbell, *American Indian Languages*, 101.

39. Campbell, *American Indian Languages*, 102.

40. *Nature*, August 26, 1922, 282.

41. "'Million-Dollar' Man-Ape's Tooth Shattered; Museum's Prize Falls from X-Ray Man's Hand."

42. Echo-Hawk, "Merciless Greetings, Wicked Servants of the Age of Archaeoracialism," 21.

43. Echo-Hawk, "Merciless Greetings, Wicked Servants of the Age of Archaeoracialism," 21, 23.

44. Echo-Hawk, "Merciless Greetings, Wicked Servants of the Age of Archaeoracialism," 22.

## Chapter 8

1. Meltzer, "Why Don't We Know When the First People Came to North America?," 30 (italics added).

2. Neimark, "Pinkie Pokes Holes in Human Evolution," 55.

3. Neimark, "Pinkie Pokes Holes in Human Evolution," 41.

4. Meltzer, *First Peoples in a New World*, 203.

5. Meltzer, *First Peoples in a New World*, 207.

6. Dillehay, "Probing Deeper into First American Studies," 974.

7. Goebel, Waters, and O'Rourke, "The Late Pleistocene Dispersal of Modern Humans in the Americas," 1497.

8. Dillehay, "Probing Deeper into First American Studies," 975.

9. Foster, *Pre-Historic Races of the United States of America*, 354, quoted in Lyell, *The Geological Evidence of the Antiquity of Man*, 338.

10. Foster, *Pre-Historic Races of the United States of America*, 355.

11. Reid, "Networking Genetics, Populations, and Race," 50.

12. "A Nomenclature System for the Tree of Human Y-Chromosomal Binary Haplogroups," US National Institutes of Health's National Library of Medicine, ncbi.nlm.nih.gov/pmc/articles/PMC155271/#!po=1.92308 (accessed July 5, 2019).

13. The exact number of haplogroups is difficult to extract from the scientific literature. In addition to haplogroups, there are also subgroups that are frequently treated as if they, not the higher-level haplogroup, are the metric for genetic distinctions. Below that is a haplotype. New haplogroups are still being discovered.

14. University of Illinois, http://www.scs.illinois.edu/~mcdonald/WorldHaplogroups-Maps.pdf (accessed July 5, 2019).

15. Chatters, *Ancient Encounters*, 234–35.

16. Chatters, *Ancient Encounters*, 207.

17. Chatters, *Ancient Encounters*, 234.

18. Chatters, *Ancient Encounters*, 235.

19. Chatters, *Ancient Encounters*, 235.

20. Chatters, *Ancient Encounters*, 236–37.

21. Chatters, *Ancient Encounters*, 237.

22. Raff, "What the Ancient DNA Discovery Tells Us About Native American Ancestry."

23. Rice, "20 Things You Didn't Know About Viruses."

24. Meltzer, "Why Don't We Know When the First People Came to North America?," 35.

25. Center for the Study of the First Americans, http://centerfirstamericans.org/who.php (accessed March 23, 2010).

26. Harmon, "Where'd You Go with My DNA?"

27. Chatters, *Ancient Encounters*, 116.

28. Deloria, "Foreword," xviii.

29. Brooks, "What Does It Mean to Be Human?," 3.

30. Wilmsen, "An Outline of Early Man Studies in the United States," 5.

31. Barton, *Observations on Some Parts of Natural History*, 89, cited in Wilmsen, "An Outline of Early Man Studies in the United States," 5.

32. Foster, *Pre-Historic Races of the United States of America*, 318.

33. Foster, *Pre-Historic Races of the United States of America*, 318.

34. Donnelly, *Atlantis*, 435.

35. Foster, *Pre-Historic Races of the United States of America*, 318, 319.

36. Foster, *Pre-Historic Races of the United States of America*, 319.

37. Edward Sapir, quoted in Meltzer, "Why Don't We Know When the First People Came to North America?," 27.

38. Greenberg, *Language in the Americas*.

39. Greenberg, *Language in the Americas*, 38.

40. For example, see Turner, "Sinodonry and Sundadonty," 72–76. Turner was prolific in his writings in the 1980s, and several other publications are contained in the bibliography.

41. Meltzer, *First Peoples*, 153.

42. Meltzer, "Why Don't We Know When the First People Came to North America?," 29.

43. Smithsonian Institution, http://anthropology.si.edu/outreach/indbibl/american-indianlanguages.pdf.

44. Adovasio, *The First Americans*, 113.

45. Adovasio, *The First Americans*, 114–16.

46. Chatters, *Ancient Encounters*, 128.

47. Adovasio, *The First Americans*, 114.

48. Goebel, Waters, and O'Rourke, "The Late Pleistocene Dispersal of Modern Humans in the Americas," 1501.

49. Goebel, Waters, and O'Rourke, "The Late Pleistocene Dispersal of Modern Humans in the Americas," 1498.

50. Fiedel, *Prehistory of the Americas*, ix.

51. Adovasio, *The First Americans*, 109.

52. Adovasio, *The First Americans*, 20.

53. Adovasio, *The First Americans*, 116–17.

54. Wilmsen, "An Outline of Early Man Studies in the United States," 16.


# Chapter 9

1. Thomas Jefferson, quoted in Adovasio, *The First Americans*, 15.

2. Adovasio, *The First Americans*, 15.

3. Adovasio, *The First Americans*, xviii.

4. Huckell and Kilby, "Late Pleistocene North America and Early Paleoindians," viii.

5. Adovasio, *The First Americans*, 76.

6. Fiedel, *Prehistory of the Americas*, 53.

7. Golinski, *Making Natural Knowledge*, 84.

8. Hacking, *Representing and Intervening*, 199.

9. Latour, "Give Me a Laboratory and I Will Raise the World," 141–70.

10. Observed by the author during a site visit on July 30, 2010.

11. Golinski, *Making Natural Knowledge*, 98.

12. Latour, "Give Me a Laboratory and I Will Raise the World," 141–70.

13. Adovasio, *The First Americans*, 88.

14. Adovasio, *The First Americans*, 87.

15. Rothenberg, "Joseph Henry."

16. Observed by the author during a site visit on July 30, 2010.

17. http://www.kochind.com/files/KochDavid.pdf (accessed February 15, 2011).

18. Observed by the author during a site visit to both museums on July 30, 2010.

19. Eleven native communities, including the Coast Salish, Makah, Nuu-chah-nulth, Kwakwaka'wakw, Heiltsuk, Nuxalk, Tsimshian, Gitxsan, Nisga'a, Haida, and Tlingit; Smithsonian National Museum of the American Indian, Office of Public Affairs, http://americanindian.si.edu/press/releases/2006-01-23_listening_to_our_ancestors_press_release_for_web.pdf (accessed February 15, 2011).

20. Conversation on July 30, 2011, between the author and a museum staff member in the NMAI Resource Center in Washington, DC.

21. Observed by the author on July 30, 2010.

22. Observed by the author on July 30, 2010.

# Chapter 10

1. Adovasio, *The First Americans*, xviii.

2. McGuire, *Archaeology as Political Action*, 235.

3. Zimmerman, "Multivocality, Descendant Communities, and Some Epistemological Shifts Forced by Repatriation," 97.

4. Bray, "Repatriation and Archaeology's Second Loss of Innocence," 79.

5. Foster, *Pre-Historic Races of the United States of America*, 95.

6. William Faulkner, *Requiem for a Nun*, cited in Bray, "Repatriation and Archaeology's Second Loss of Innocence," 79.

7. Feyerabend, *Against Method*, 3.

8. Longino, *Science as Social Knowledge*, 15.

9. Longino, *Science as Social Knowledge*, 57.

10. Killion, *Opening Archaeology*, iii–iv.

11. Killion, *Opening Archaeology*, iii–iv.

12. McManamon, "The Native American Graves Protection and Repatriation Act and First Americans Research," 145.

13. Tierney, "Social Scientist Sees Bias Within."

14. Twain, *A Connecticut Yankee in King Arthur's Court*, 468.

15. Pickering, *The Mangle of Practice*, 187.

16. Merton, *The Sociology of Science*, 252.

17. Foucault, *Madness and Civilization*, 94.

18. Adovasio, *The First Americans*, 23.

19. Anthropological Society of London, *Memoirs Read Before the Anthropological Society of London, 1863–1864*.

20. Anthropological Society of London, *Memoirs Read Before the Anthropological Society of London, 1863–1864*, 72–119.

21. Anthropological Society of London, *Memoirs Read Before the Anthropological Society of London, 1863–1864*, 72.

22. Deloria, *Red Earth, White Lies*, 83–84.

23. For example, see Epstein, "The Construction of Lay Expertise," 408–37; Collins and Pinch, "The Science of the Lamb," 113–25.

24. Galtung, "Scientific Colonialism," 10–15.

25. Zimmerman, "Multivocality, Descendant Communities, and Some Epistemological Shifts Forced by Repatriation," 91.

26. McGuire, *Archaeology as Political Action*, 24.

27. Bray, "Repatriation and Archaeology's Second Loss of Innocence," 81.

28. Gullapalli, "Who Owns Antiquity," 48.

29. McGuire, *Archaeology as Political Action*, 23.

30. McGuire, *Archaeology as Political Action*, 18.

31. Thomas, "American Archaeology in the Twenty-First Century," 58.

32. Morrow and Morrow, "Geographic Variation in Fluted Projectile Points," 147.

33. Morrow and Morrow, "Geographic Variation in Fluted Projectile Points," 157.

34. Meltzer, "Why Don't We Know When the First People Came to North America?," 26.

35. Thomas, "American Archaeology in the Twenty-First Century," 74.

36. McGuire, *Archaeology as Political Action*, 29.

37. Zimmerman, "Multivocality, Descendant Communities, and Some Epistemological Shifts Forced by Repatriation," 94.

38. McGuire, *Archaeology as Political Action*, 123.

39. Stapp, "The Impact of NAGPRA on Anthropology," 212.

40. Killick and Goldberg, "A Quiet Crisis in American Archaeology," 8.

41. Watkins, "The Repatriation Arena," 162.

42. Clark, "Should One Size Fit All?," 39.


# Chapter 11

1. Deloria, *Red Earth, White Lies*, 68–69, cited in Weaver, "Indian Presence with No Indians Present," 24.

2. Johnson, "Indian Tribes' Creationists Thwart Archeologists," cited in Weaver, "Indian Presence with No Indians Present," 19.

3. Curtin. *Creation Myths of Primitive America*, xxxi.

4. Paul Chaat Smith exhibit plaque, Museum of the American Indian, Smithsonian Institution, observed by the author on July 29, 2010.

5. Golinski, *Making Natural Knowledge*, 21.

6. United States House of Representatives, "Providing for the Protection of Native American Graves, and for other purposes."

7. Bray, "Repatriation and Archaeology's Second Loss of Innocence," 79.

8. Feyerabend, *Against Method*, 32.

9. The Native American Graves Protection and Repatriation Act, 25 U.S.C. 3005 a(4), as amended.

10. Richard Feynman interview in 1981, cited in Robbins, *The Pleasure of Finding Things Out*, 186–97.

11. Archaeological Institute of America, "Who Owns the Dead?," 16.

12. Weaver, "Indian Presence with No Indians Present," 14.

13. Archaeological Institute of America, "Who Owns the Dead?," 58.

14. Killion, *Opening Archaeology*, 4.

15. Killion, *Opening Archaeology*, 5.

16. Killion, *Opening Archaeology*, 10.

17. Antiquities Act of 1906.

18. Archaeological Resources Protection Act of 1979.

19. United States Senate, S. R. 2843.

20. United States House of Representatives, "Report 4027."

21. Archaeological Institute of America, "Who Owns the Dead?," 58.

22. Archaeological Institute of America, "Who Owns the Dead?," 58.

23. United States Department of the Interior, *Federal Agencies Task Force Report, American Indian Religious Freedom Act Report,* 64, cited in Weaver, "Indian Presence with No Indians Present," 15.

24. United States House of Representatives, "Providing for the Protection of Native American Graves, and For Other Purposes," 9.

25. United States House of Representatives, "Providing for the Protection of Native American Graves, and For Other Purposes," 13.

26. Walter and Roger Echo-Hawk, cited in Weaver, "Indian Presence with No Indians Present," 17.

27. Owsley. "From Jamestown to Kennewick," 138.

28. Weaver, "Indian Presence with No Indians Present," 17.

29. McManamon, "The Native American Graves Protection and Repatriation Act and First Americans Research," 144.

30. Kintigh. "Repatriation as a Force of Change in Southwestern Archaeology," 203.

31. *Bonnichsen v. United States of America,* 23, 24.

32. Killion, *Opening Archaeology,* 12.

33. Watkins, "The Repatriation Arena," 164–65.

34. Observed by the author on July 29, 2010, at the Museum of the American Indian, Smithsonian Institution.

35. Johnson, "Indian Tribes' Creationists Thwart Archeologists."

36. Weaver, "Indian Presence with No Indians Present," 19.

37. Johnson, "Indian Tribes' Creationists Thwart Archeologists," cited in Weaver, "Indian Presence with No Indians Present," 20.

38. Gould, *Leonardo's Mountain of Clams and the Diet of Worms,* 269–84.

39. Egan, "Tribe Stops Study of Bones that Challenge History," cited in Weaver, "Indian Presence with No Indians Present," 19.

40. Deloria, *Red Earth, White Lies,* 24.

41. Weaver, "Indian Presence with No Indians Present," 19.

42. McGuire, *Archaeology as Political Action,* xii.

43. Killion, *Opening Archaeology,* 4.

## Chapter 12

1. Chatters, *Ancient Encounters,* 20.

2. Chatters, *Ancient Encounters,* 21.

3. Chatters, *Ancient Encounters,* 27.

4. Citizendium: The Citizens Compendium, "Kennewick Man Bibliography."

5. Chatters, *Ancient Encounters,* 53.

6. Chatters, *Ancient Encounters,* 63.

7. Thomas, *Skull Wars,* Prologue, xxi.

8. *Bonnichsen v. United States of America*, 3, footnote 3.

9. Chatters, *Ancient Encounters*, 55.

10. Chatters, *Ancient Encounters*, 53.

11. A National Park Service website that provides a detailed listing and links to the many studies is located at http://www.nps.gov/archeology/kennewick/index.htm (accessed April 30, 2011).

12. The Confederated Tribes of the Umatilla Indian Reservation are an association representing approximately 3,100 members of the Umatilla, Walla Walla, and Cayuse tribes, http://www.umatilla.nsn.us (accessed April 30, 2011).

13. Quoted in *New York Times*, September 30, 1996, http://www.nytimes.com/1996/09/30/us/tribe-stops-study-of-bones-that-challenge-history.html (accessed April 29, 2011).

14. Chatters, *Ancient Encounters*, 87.

15. McManamon, "The Native American Graves Protection and Repatriation Act and First Americans Research," 147.

16. Chatters, *Ancient Encounters*, 110.

17. Meltzer, *First Peoples in a New World*, 178.

18. Chatters, *Ancient Encounters*, 90.

19. Chatters, *Ancient Encounters*, 96.

20. Chatters, *Ancient Encounters*, 103.

21. Chatters, *Ancient Encounters*, 101.

22. *New York Times*, September 30, 1996, http://www.nytimes.com/1996/09/30/us/tribe-stops-study-of-bones-that-challenge-history.html (accessed April 29, 2011).

23. Chatters, *Ancient Encounters*, 91.

24. Chatters, *Ancient Encounters*, 91.

25. Thomas, *Skull Wars*, 243.

26. *Bonnichsen v. US*, 5.

27. *Bonnichsen v. US*, 31.

28. The Asatru Folk Assembly is a membership organization headquartered in California, http://www.runestone.org (accessed April 30, 2011).

29. *Bonnichsen v. US*, 6, footnote 10.

30. Chatters, *Ancient Encounters*, 93.

31. Weaver, "Indian Presence with No Indians Present," 18–19.

32. *Bonnichsen v. US*, 14.

33. *Bonnichsen v. US*, 14, footnote 24.

34. Chatters, *Ancient Encounters*, 62.

35. *Bonnichsen v. US*, 8.

36. Chatters, *Ancient Encounters*, 94.

37. July 15, 2005, letter to Senator John McCain, Chairman, Senate Committee on Indian Affairs, from E. J. Neiburger, Director, Ethnic Minority Council of America, http://www.friendsofpast.org/nagpra/news/EMCLetter.pdf (accessed April 12, 2011).

38. Chatters, *Ancient Encounters*, 171.

39. Chatters, *Ancient Encounters*, 105.

40. Adovasio, *First Americans*, 4.

41. Thomas, *Skull Wars*, 235.

42. National Park Service Archaeology Program, http://www.nps.gov/archeology/kennewick/index.htm (accessed September 2, 2019).

43. Burke Museum of Natural History and Culture, http://www.washington.edu/burkemuseum/kman (accessed April 30, 2011).

44. Meltzer, *First Peoples in a New World*, 180.

45. Li et al., "Evidence that a West-East Admixed Population Lived in the Tarim Basin as Early as the Early Bronze Age," 1.

46. Pringle, "Battle for the Xinjiang Mummies," 34.

47. Pringle, "Battle for the Xinjiang Mummies," 30.

48. Pringle, "Battle for the Xinjiang Mummies," 30.

49. Mair, "Ancient Mummies of the Tarim Basin."

50. Over the past twenty years, Mair's teams have included Paolo Francalacci, a geneticist at the University of Sassari in Italy; Idelisi Abuduresule, formerly director of the Institute of Archaeology and Cultural Relics in Ürümchi, Xinjiang, China; and Elizabeth W. Barber of Occidental College

51. Pringle, "Battle for the Xinjiang Mummies," 34.

52. Pringle, "Battle for the Xinjiang Mummies," 35.

53. Li et al., "Evidence that a West-East Admixed Population Lived in the Tarim Basin as Early as the Early Bronze Age," 9, 5–6.

54. Li et al., "Evidence that a West-East Admixed Population Lived in the Tarim Basin as Early as the Early Bronze Age," 2.

55. Kennicott, "Following Controversy, Mummies at Penn Museum Remain Objects of Mystery."

56. Rothstein, "Another Stop on a Long, Improbable Journey."

## Conclusions

1. Adovasio, *The First Americans*, 4.

2. Respectively, these are descriptions of revolutionary scientific change by Kuhn, *The Structure of Scientific Revolutions*; Bachelard, *The Formation of the Scientific Mind*; Eldredge and Gould, "Punctuated Equilibria," 82–115; and Foucault, *The Order of Things*.

3. Laudan, *Progress and Its Problems*, 79.

4. Longino, *Science as Social Knowledge*, 48.

5. Ioannidis, "Why Most Published Research Findings Are False," 696.

6. Freeman, quoted in July/August 2010 issue of *Discover*, 56, an excerpt from *Wrong*.

7. Feynman, "The Development of Space-Time View of Quantum Electrodynamics," 155.

8. Observation by the author, Mountain Creek Lodge, Pipestem State Park, West Virginia, September 17, 2011.

# BIBLIOGRAPHY

Abbott, Charles Conrad. "Paleolithic Man: A Last Word." *Science* 20, no. 515 (1892): 344–45.

———. "Palaeolithic Man in North America." *Science* 20, no. 510 (1892): 270–71.

Adair, James. *The History of the American Indians.* London: Edward and Charles Dilly, 1777.

Adovasio, James M. *The First Americans: In Pursuit of Archaeology's Greatest Mystery. With Jake Page.* New York: The Modern Library, 2003.

Adovasio, James M., and Ronald C. Carlisle. "Some Thoughts on Cultural Resource Management Archaeology in the United States." *Antiquity* 62 (1988): 72–87.

Adovasio, J. M. "Meadowcroft Rockshelter." In *Early Man in the New World from a Circum-Pacific Perspective,* edited by Alan Lyle Bryan. Edmonton: University of Alberta, 1979.

Adovasio, J. M., J. Donahue, and R. Stuckenrath. "The Meadowcroft Rockshelter Radiocarbon Chronology 1975–1990." *American Antiquity* 55, no. 2 (1990): 348–54.

Adovasio, J. M., J. Donahue, R. C. Carlisle, K. Cushman, R. Stuckenrath, and P. Wiegman. "Meadowcroft Rockshelter and the Pleistocene/Holocene Transition in Southwestern Pennsylvania." In *Contributions in Quaternary Vertebrate Paleontology: A Volume in Memorial to John E. Guilday,* edited by Hugh H. Genoways and Mary R. Dawson, 347–69. Pittsburgh: Carnegie Museum of Natural History, 1984.

Adovasio, J. M., J. D. Gunn, J. Donahue, and R. Stuckenrath. "Excavations at Meadowcroft Rockshelter, 1973–1974: A Progress Report." *Pennsylvania Archaeologist* 45, no. 3 (1975): 1–30.

———. "Meadowcroft Rockshelter, 1977: An Overview." In *Readings in Late Pleistocene North America and Early Paleoindians: Selections from American Antiquity,* edited by Bruce B. Huckell and J. David Kilby, 261. Washington, DC: Society for American Archaeology, 2004.

———. "Meadowcroft Rockshelter: Retrospect 1976." *Pennsylvania Archaeologist* 47, no. 2–3 (1977): 1–93.

———. "Progress Report on the Meadowcroft Rockshelter: A 16,000-Year Chronicle." In *Amerinds and Their Paleoenvironments in North America,* edited by Walter S. Newman and Bert Saiwen, 137–60. New York: New York Academy of Sciences, 1977.

Adovasio, J. M., and D. R. Pedler. "Monte Verde and the Antiquity of Humankind in the Americas." *Antiquity* 71, no. 273 (1997): 573–80.

American Anthropological Association. *"AAA Responds to Public Controversy over Science in Anthropology."* http://www.aaanet.org/issues/press/AAA-Responds-to-Public-Controversy-Over-Science-in-Anthropology.cfm.

———. *"Connect with AAA."* Accessed August 10, 2019. https://www.americananthro.org/ConnectWithAAA/Content.aspx?ItemNumber=1650&navItemNumber=760.

American Association for the Advancement of Science. "Five Pre-Clovis Sites That Won't Go Away." *Science* 249, no. 4970 (August 17, 1990): 740.

American Association of Physical Anthropologists (AAPA). "AAPA Statement on Biological Aspects of Race." *American Journal of Physical Anthropology* 101 (1996): 569–70.

Anawak, Jack. "Inuit Perceptions of the Past." In *Who Needs the Past? Indigenous Values and Archaeology*, edited by R. Layton, 45–50. London: Unwin Hyman, 1989.

Anderson, David G., and J. Christopher Gillam. "Paleoindian Colonization of the Americas: Implications from an Examination of Physiography, Demography, and Artifact Distribution." In *Readings in Late Pleistocene North America and Early Paleoindians: Selections from American Antiquity*, edited by Bruce B. Huckell and J. David Kilby, 50. Article of the same title also originally published in *American Antiquity* 65, no. 1 (2000): 44.

Anthropological Society of London. *Memoirs Read Before the Anthropological Society of London, 1863–1864.* Vol. 1. London: Trübner and Co., 1865.

Antiquities Act of 1906. 16 U.S.C. 431-433. Accessed January 26, 2021. http://www.nps.gov/history/local-law/anti1906.htm.

Antropova, V. "The Koryaks." In *The Peoples of Siberia*, edited by M. G. Levin and L. P. Potapov, 851–75. Chicago: University of Chicago Press, 1964.

Archaeological Institute of America. "Who Owns the Dead? A Controversial Amendment to the Federal Repatriation Law Complicates the Relationship between Native Americans and Archaeologists." *Archaeology* 64, no. 1 (January/February 2011): 16.

Archaeological Resources Protection Act of 1979. 16 U.S.C. 470aa–mm. Accessed January 26, 2021. http://www.nps.gov/history/local-law/fhpl_archrsrcsprot.pdf.

Arnold, Thomas G. *"The Ice-Free Corridor: Biogeographical Highway or Environmental Cul-de-Sac."* PhD diss., Simon Fraser University, 2006.

Asatru Folk Assembly. Accessed January 26, 2021. http://www.runestone.org.

Atwater, Caleb. *Transaction from American Antiquarian Society in Boston. 1820. Later self-published as Writings of Caleb Atwater.* Columbus: Author, 1833. Modern reprint available through Applewood Books, Carlisle, Massachusetts.

Bachelard, Gaston. *The Formation of the Scientific Mind: A Contribution to a Psycho-analysis of Objective Knowledge.* Manchester, UK: Clinamen Press, 2002.

Banning, Edward Bruce. *The Archaeologist's Laboratory: The Analysis of Archaeological Data.* New York: Springer, 2002.

Barnard, Alan. *History and Theory in Anthropology.* Cambridge, UK: Cambridge University Press, 2000.

Barton, Benjamin Smith. *New Views of the Origin of the Tribes and Nations of America.* 2nd ed. Philadelphia: John Bioren, 1798.

———. *Observations on Some Parts of Natural History: To Which Is Prefixed an Account of Several Remarkable Vestiges of an Ancient Date, Which Have Been Discovered in Different Parts of North America*. Part I. London: Charles Dilly, 1787.

Barton, C. Michael, Geoffrey A. Clark, David R. Yesner, and Georges A. Pearson, eds. *The Settlement of the American Continents: A Multidisciplinary Approach to Human Biogeography*. Tucson: University of Arizona Press, 2004.

Beals, Ralph L. "Father Acosta on the First Peopling of the New World." *American Antiquity* 23, no. 2 (1957): 182–83.

Beauchamp, William M. *The Iroquois Trail, or Foot-Prints of the Six Nations, in Customs, Traditions, and History, in Which Are Included David Cusick's Sketches of Ancient History of the Six Nations*. Fayetteville, NY: Author, 1891.

Behling Hermann. "First Report on New Evidence for the Occurrence of Podocarpus and Possible Human Presence at the Mouth of the Amazon during the Late Glacial." *Vegetation History and Archaeobotany* 5 (1996): 241–46.

Bell, James. "On Capturing Agency in Theories about Prehistory." In *Representations in Archaeology*, edited by Jean-Claude Gardin and Christopher S. Peebles, 30–55. Bloomington: Indiana University Press, 1992.

Bentley, R. Alexander, Matthew W. Hahn, and Stephen J. Shennan. "Random Drift and Culture Change." *Proceedings of the Royal Society* B 271 (2004): 1443–50.

Berrett, Dan. "*Anthropology Without Science*." *Inside Higher Ed*, November 30, 2010. https://www.insidehighered.com/news/2010/11/30/anthropology-without-science.

Biagioli, Mario. *Galileo, Courtier: The Practice of Science in the Culture of Absolutism*. Chicago: University of Chicago Press, 1993.

Bieder, Robert E. *Science Encounters the Indian, 1820–1880: The Early Years of American Ethnology*. Norman: University of Oklahoma Press, 1986.

Binford, Lewis R. "Archaeology as Anthropology." *American Antiquity* 28, no. 2 (1962): 217–25.

Binford, Sally, and Lewis R. Binford, eds. *New Perspectives in Archaeology*. Chicago: Aldine, 1968.

Biolsi, Thomas, and Larry Zimmerman, eds. *Indians and Anthropologists: Vine Deloria, Jr., and the Critique of Anthropology*. Tucson: University of Arizona Press, 1997.

Birdsell, J. B. "The Problem of the Early Peopling of the Americas as Viewed from Asia." In *The Physical Anthropology of the American Indian*, edited by W. S. Laughlin, 1–68. Ann Arbor, MI: Edwards Brothers, 1951.

Blackhawk, N. "Reconciliation of Contested Truths." *Anthropology Newsletter* 38, no. 3 (1997): 2.

Blakey, Michael L. "American Nationality and Ethnicity in the Depicted Past." In *The Politics of the Past*, edited by Peter Gatherole and David Lowenthal, 38–48. London: Unwin Hyman, 1990.

Boas, Franz. *Race, Language and Culture*. New York: Macmillan Company, 1940. First published in 1899 in London.

Boldurian, Anthony T., and John L. Cotter. *Clovis Revisited: New Perspectives on Paleoindian Adaptations from Blackwater Draw, New Mexico.* Philadelphia: University Museum, University of Pennsylvania, 1999.

Bonatto, S. L., A. J. Redd, F. M. Salzano, and M. Stoneking. "Lack of Ancient Polynesian-American Contact." *American Journal of Human Genetics* 59, no. 1 (July 1996): 253–58.

Bonatto, S. L., and F. M. Salzano. "A Single and Early Origin for the Peopling of the Americas Supported by Mitochondrial DNA Sequence Data." *Proceedings of the National Academy of Sciences* 94 (1997): 1866–71.

———. "Diversity and Age of the Four Major mtDNA Haplogroups, and Their Implications for the Peopling of the New World." *American Journal of Human Genetics* 61 (1997): 1413–23.

*Bonnichsen v. United States of America*, US District Court for the District of Oregon, Opinion and Order Civil No. 96-1481-JE, 23.

Bonnichsen, R., and D. Young. "Early Technological Repertoires: Bone to Stone." *Canadian Journal of Anthropology* 1 (1980): 123–128.

———. "Preface." In *Early Man in America*, edited by A. Bryan, v–vi. Edmonton: University of Alberta, 1978.

Bonnichsen, Robson, ed. *Who Were the First Americans: Proceedings of the 58th Annual Biology Colloquium.* Corvallis: Center for the Study of the First Americans, Oregon State University, 1999.

Bonnichsen, Robson, and D. Gentry Steele, eds. *Method and Theory for Investigating the Peopling of the Americas.* Corvallis: Center for the Study of the First Americans, Oregon State University, 1994.

Bonnichsen, Robson, and Karen L. Turnmire, eds. *Clovis: Origins and Adaptations.* Corvallis: Center for the Study of the First Americans, Oregon State University, 1991.

———. *Ice Age People of North America: Environments, Origins, and Adaptations of the First Americans.* Corvallis: Center for the Study of the First Americans, Oregon State University, 1999.

Bonnichsen Robson, Bradley T. Lepper, Dennis Stanford, and Michael R. Waters, eds. *Paleoamerican Origins: Beyond Clovis.* College Station: Texas A&M University Press, 2005.

Bonnichsen, Robson, James L. Fastook, and Dennis Stanford. "Environmental Change and Developmental History of Human Adaptive Patterns: The Paleoindian Case." In *The Geology of North America. Vol. K-3, North America and Adjacent Oceans during the Last Deglaciation*, edited by W. F. Ruddiman and H. E. Wright Jr., 403–24. Boulder, CO: Geological Society of America, 1987.

Borrero, Luis Alberto. "Paleoindians without Mammoths and Archaeologists without Projectile Points? The Archaeology of the First Inhabitants of the Americas." In *Paleoindian Archaeology: A Hemispherical Perspective*, edited by Juliet E. Morrow and Cristobál Gnecco, 9–20. Gainesville: University Press of Florida, 2006.

Bortolini, Maria-Catira, Francisco M. Salzano, Mark G. Thomas, Steven Stuart, Selja P. K. Nasanen, Claiton H. D. Bau, Mara H. Hutz, et al. "Y-chromosome Evidence for Differing Ancient Demographic Histories in the Americas." *American Journal of Human Genetics* 73 (2003): 524–39.

Bowdler, Sandra. "Peopling Australasia: the 'Coastal Colonization' Hypothesis Re-Examined." In *The Emergence of Modern Humans*, edited by Paul Mellars, 327–43. Edinburgh: Edinburgh University Press, 1990.

Bowers Museum. "*Secrets of the Silk Road*" exhibit. https://www.bowers.org/index.php/past-exhibition/secrets-of-the-silk-road.

Boyd, Julian P. *The Papers of Thomas Jefferson*. Vol. 12, no. 7 (August 1787 to March 1788). Princeton, NJ: Princeton University Press, 1955.

Brace, C. Loring. "The Roots of the Race Concept in American Physical Anthropology." In *A History of American Physical Anthropology, 1930–1980*, edited by F. Spencer, 11–30. New York: Academic Press, 1982.

Brace, C. Loring, Noriko Segochi, and Mary L. Brace. "Exploring the Kennewick Connection." In *Kennewick Man: Perspectives on the Ancient One*, edited by Heather Burke, Claire Smith, Dorothy Lippert, Joe Watkins, and Larry Zimmerman, 153–68. Walnut Creek, CA: Left Coast Press, 2008.

Brace, C. Loring, A. Russell Nelson, Noriko Seguchi, Hiroaki Oe, Leslie Sering, Pan Qifeng, Li Yongyi, and Dashtseveg Tumen. "Old World Sources of the First New World Human Inhabitants: A Comparative Craniofacial View." *Proceedings of the National Academy of Sciences USA* 98 (2001): 10017–22.

Bradley, B., and D. Stanford. "The North Atlantic Ice-Edge Corridor: A Possible Paleolithic Route to the New World." *World Archaeology* 36 (2004): 459–78.

Bradley, James W., and John Bryant. "NAGPRA's Implications." *Science* 264, no. 5161 (May 13, 1994): 890.

Bray, Tamara L. "Repatriation, Power Relations and the Politics of the Past." *Antiquity* 70, no. 268 (1996): 440–44.

———. "Repatriation and Archaeology's Second Loss of Innocence: On Knowledge, Power, and the Past." In *Opening Archaeology: Repatriation's Impact on Contemporary Research and Practice*, edited by Thomas W. Killion, 79. Santa Fe: School for Advanced Research Press, 2008.

Bray, Tamara L., ed. *The Future of the Past: Archaeologists, Native Americans, and Repatriation*. New York: Garland Publishing, 2001.

Bray, Warwick. "The Paleoindian Debate." *Nature* 332 (1988): 107.

Brerewood, Edward. *Enquiries Touching the Diversity of Languages and Religions, through the Chief Parts of the World*. 2nd ed. London: Samuel Mearne, John Martyn, and Henry Herringman, 1674. First edition, 1614.

Brew, John Otis. Review of *Man and the Glacial Period* by G. Frederick Wright. *Science* 20, no. 5081 (1894): 249.

Britain, Marcus, and Timothy Clack, eds. *Archaeology and the Media*. London: University College London Press, 2005.

Brooks, Alison S. "What Does it Mean to be Human? A Behavioral Perspective." *AnthroNotes: Museum of Natural History Publication for Educators* 31, no. 1 (Spring 2010): 3.

Browman, David L., and Stephen Williams. *New Perspectives on the Origins of Americanist Archaeology. Tuscaloosa*: University of Alabama Press, 2002.

Brown, B. "Recent Finds Relating to Prehistoric Man in America." *Bulletin of the New York Academy of Medicine* 4 (1928): 824–28.

Brown, Michael F. *Who Owns Native Culture?* Cambridge, MA: Harvard University Press, 2003.

Bruning, Susan B. "Complex Legal Legacies: The Native American Graves Protection and Repatriation Act, Scientific Study, and Kennewick Man." *American Antiquity* 71, no. 3 (2006): 501–21.

Bryan, Alan L., and Ruth Gruhn. "Some Difficulties in Modeling the Original Peopling of the Americas." *Quaternary International* 109–10 (2003): 175–79.

Bryan, Alan Lyle. "An Overview of Paleo-American Prehistory from a Circum-Pacific Perspective." In *Early Man in America from a Circum-Pacific Perspective*, edited by Alan Lyle Bryan, 306–27. Edmonton: University of Alberta, 1978.

———. Discussion of *Folsom Culture and Its Age* by H. Brown. *Geological Society of America Bulletin* 40 (1986): 128–29.

Bryan, Alan Lyle, ed. *Early Man in America from a Circum-Pacific Perspective*. Edmonton: University of Alberta, 1978.

———. *New Evidence for the Pleistocene Peopling of the Americas*. Orono: Center for the Study of Early Man, University of Maine, 1986.

Bryant, David, Flavia Filimon, and Russel D. Gray. "Untangling Our Past: Languages, Trees, Splits, and Networks." In *The Evolution of Cultural Diversity: A Phylogenetic Approach*, edited by Ruth Mace, Clare J. Holden, and Stephen J. Shennan, 67–83. London: University College London Press, 2005.

Bryant, Jacob. *A New System, or an Analysis of Ancient Mythology: Wherein an Attempt Is Made to Divest Tradition of Fable and to Reduce the Truth to Its Original Purity.* 3 vols. London: T. Payne, P. Elmsly, B. White, and J. Walter, 1774.

Buchanan, Briggs, Mark Collard, and Kevan Edinborough. "Paleoindian Demography and the Extraterrestrial Impact Hypothesis." *Proceedings of the National Academy of Sciences* 105 (2008): 11651–54.

Buikstra, Jane E. "Repatriation and Bioarchaeology: Challenges and Opportunities." In *Bioarchaeology: The Contextual Analysis of Human Remains*, edited by Jane E. Buikstra and Lane A. Beck, 389–415. New York: Academic Press, 2006.

Buikstra, Jane E., and Lane A. Beck, eds. *Bioarchaeology: The Contextual Analysis of Human Remains*. New York: Academic Press, 2006.

Buikstra, Jane E., and Douglas H. Ubelaker. *Standards for Data Collection from Human Skeletal Remains*. Fayetteville: Arkansas Archeological Survey Research Series No. 44, 1994.

Burgaleta, Claudio. *Jose de Acosta, S.J. (1540–1600): His Life and Thought*. Chicago: Loyola Press, 1999.

Burke Museum of Natural History and Culture. https://www.burkemuseum.org/.

Campbell, Lyle. *American Indian Languages: The Historical Linguistics of Native America*. Oxford, UK: Oxford University Press, 1997.

———. Comment on *The Settlement of the Americas: A Comparison of the Linguistic, Dental, and Genetic Evidence* by Joseph H. Greenberg, Christy G. Turner III, and Stephen L. Zegura. *Current Anthropology* 27 (1986): 490–91.

———. Review of *Languages in the Americas* by Joseph H. Greenberg. *Language* 64 (1988): 591–615.

Carlisle, Ronald C., ed. *Americans Before Columbus: Ice Age Origins*. Pittsburg: University of Pittsburgh, 1988.

Carver, Jonathan. *Travels through the Interior Parts of North America (1766–1768)*. London: J. Walter, 1778.

Center for the Study of the First Americans. https://liberalarts.tamu.edu/csfa/.

Chambers, Robert. *Explanations: A Sequel to "Vestiges of the Natural History of Creation."* London: J. Churchill, 1844.

———. *Vestiges of the Natural History of Creation*. London: J. Churchill, 1844.

Chatters, James C. *Ancient Encounters: Kennewick Man and the First Americans*. New York: Touchstone, 2001.

———. "The Recovery and First Analysis of an Early Holocene Human Skeleton from Kennewick, Washington." *American Antiquity* 65, no. 2 (2000): 291–316.

Chief Joseph. "My Son, Stop Your Ears." In *1879 Testimony Before Congress, in Native American Testimony: A Chronicle of Indian-White Relations from Prophecy to the Present, 1492–1992*, edited by Peter Nabokov, 129–33. New York: Viking, 1991.

Childe, V. Gordon. "Is Prehistory Practical?" *Antiquity* 7 (1933): 410–18.

Chinard, Gilbert. "Jefferson and the American Philosophical Society." *Proceedings of the American Philosophical Society* 87, no. 3, Bicentennial of Thomas Jefferson (July 14, 1943): 263–76.

Citizendium: The Citizens Compendium. "Kennewick Man Bibliography." Accessed January 26, 2021. http://en.citizendium.org/wiki/Kennewick_Man/Bibliography.

Clark, Geoffrey. "Should One Size Fit All? Some Observations on Killick and Goldberg." *SAA Archaeological Record* 10, no. 1 (January 2010): 39–42.

Clovis News Journal. http://www.cnjonline.com.php56-17.ord1-1.websitetestlink.com/news/.

Cohen, J., and N. Swidler. "Painting a New Face on CRM: Integrating Traditional Culture and Archaeology." *SAA Bulletin* 15, no. 1 (1997): 24–25.

Colbert, E. H. *"The Association of Man with Extinct Mammals in the Western Hemisphere."* Proceedings of the Eighth American Scientific Congress held in Washington May 10–18, 1940, under the Auspices of the Government of the United States of America . . ., edited by Paul H. Oehser, 17–29. Washington, DC: Department of State, 1941-43.

Collins, Harry, and Trevor Pinch, "The Science of the Lamb: Chernobyl and the Cumbrian Sheep Farmers." In *The Golem at Large: What you Should Know about Technology, 113–25*. Cambridge, UK: Cambridge University Press, 1998.

Confederated Tribes of the Umatilla Indian Reservation. https://ctuir.org/.

Cook, Harold J. "Glacial Age Man in New Mexico." *Scientific American* 139 (1928): 38–40.

Cope, Edward D. "The Antiquity of Man in North America." *American Naturalist* 29, no. 342 (1895): 593–99.

Cuno, James. *Who Owns Antiquity? Museums and the Battle over Our Ancient Heritage.* Princeton, NJ: Princeton University Press, 2008.

Curry County, New Mexico, Chamber of Commerce. https://www.clovisnm.org/.

Curtin, Jeremiah. *Creation Myths of Primitive America.* New York: Benjamin Blom, *1969.* First published as *Creation Myths of Primitive America: In Relation to the Religious History and Mental Development of Mankind.* New York: Little, Brown, and Company, 1898.

Dalibard, Jacques. "The Canadian Experience in Heritage Preservation." In *The Challenge to Our Cultural Heritage: Why Preserve the Past?* edited by R. Yudhishthir Isar, 183–94. Washington, DC: Smithsonian Institution Press, 1986.

d'Anghiera, Pietro Martire. *De Orbe Novo: The Eight Decades of Peter Martyr d'Anghera.* 2 vols. Translated and edited by Francis Augustus MacNutt. New York: Putnam, 1912.

Daston, Lorraine, and Peter Galison. *Objectivity.* New York: Zone Books, 2007.

Davis, Hester A. "The Crises in American Archeology." *Science* 175 (1972): 267–72.

———. "Is an Archaeological Site Important to Science or to the Public, and Is There a Difference?" In *Heritage Interpretation: The Natural and Built Environment,* vol. 1, edited by David L. Uzzell, 96–99. London: Belhaven, 1989.

de Acosta, José. *Historia natural y moral de las Indias (The Natural and Moral History of the Indies).* Seville: Casa Juan de Leon, 1590.

Deloria, Vine, Jr. "Anthros, Indians, and Planetary Reality." In *Indians and Anthropologists: Vine Deloria, Jr., and the Critique of Anthropology,* edited by Thomas Biolsi and Larry J. Zimmerman, 209–21. Tucson: University of Arizona Press, 1997.

———. *Custer Died for Your Sins: An Indian Manifesto.* New York: Avon, 1969.

———. "Foreword." In *Skull Wars: Kennewick, Archaeology, and the Battle for Native American Identity,* edited by David Thomas Hurst, xvii. New York: Basic Books, 2000.

———. *God Is Red.* New York: Delta, 1973.

———. "Indians, Archaeologists, and the Future." *American Antiquity* 57, no. (1992): 595–98.

———. *Red Earth, White Lies: Native Americans and the Myth of Scientific Fact.* New York: Scribner, 1995.

Deloria, Vine, Jr., and Clifford M. Lytle. *The Nations Within: The Past and Future of American Indian Sovereignty.* Austin: University of Texas Press, 1984.

Diamond, J. "Who Were the First Americans?" *Nature* 329 (1987): 580–81.

Diaz-Andreu, Margarita, and Timothy C. Champion, eds. *Nationalism and Archaeology in Europe.* London: University College London Press, 1996.

Diebold, A. R. "Linguistic Ways to Prehistory." In *Proto Indo-European: The Archaeology of a Linguistic Problem,* edited by Susan Nacev Skomal and Edgar C. Polomé, 19–71. Washington, DC: Institute for the Study of Man, 1987.

Dikov, N. N. "The Paleolithic of Kamchatka and Chukotka and the Problem of the Peopling of America." In *Anthropology in the Pacific Rim*, edited by W. W. Fitzhugh and V. Chausonnet, 87–95. Washington, DC: Smithsonian Institution Press, 1994.

Dillehay, Tom D. "*The Battle of Monte Verde*." *The Sciences* (January/February 1997): 32.

———. "How New is the New World?" *Antiquity* 62 (1988): 94–97.

———. "Probing Deeper into First American Studies." *Proceedings of the National Academy of Sciences* 106 (2009): 971–78.

———. "A Regional Perspective of Preceramic Times in the Central Andes." *Reviews in Anthropology* 12 (1985): 193–205.

———. *The Settlement of the Americas: A New Prehistory*. New York: Basic Books, 2000.

Dillehay, Tom D., ed. *Monte Verde: A Late Pleistocene Settlement in Chile. Vol. 1, Paleoenvironment and Site Context*. Washington, DC: Smithsonian Institution Press, 1989.

———, ed. *Monte Verde: A Late Pleistocene Settlement in Chile. Vol. 2, The Archeological Context and Interpretation*. Washington, DC: Smithsonian Institution Press, 1997.

Dillehay, Tom D., and David Meltzer, eds. *The First Americans: Search and Research*, Boca Raton: CRC Press, 1991.

Dillehay, Tom D., and Michael B. Collins. "Early Cultural Evidence from Monte Verde in Chile." *Nature* 332 (1988): 150–52.

Dillehay, Tom D., Gerardo Ardila Calderón, Gustavo Politis, and Maria de Conceicao de Moraes Coutinho Beltrão. "Earliest Hunters and Gatherers of South America." *Journal of World Prehistory* 6, no. 2 (1992): 145–204.

Dillehay, Tom D., Michael B. Collins, Mario Pino, Jack Rossen, Jim Adovasio, Carlos Ocampo, Ximena Navarro, et al. "*On Monte Verde: Fiedel's Confusions and Misrepresentations*." Accessed June 26, 2011. www.uky.edu/Projects/MonteVerde/monteverde.pdf.

Dincauze, D. "An Archaeological Evaluation of the Case for Pre-Clovis Occupations." In *Advances in World Archaeology*, vol. 3, edited by Fred Wendorf and Angela E. Close, 275–323. New York: Academic Press, 1984.

Dincauze, D. F. "The Meadowcroft Papers." *Quarterly Review of Archaeology* 2 (1981): 3–4.

Dixon, E. James. *Bones, Boats, and Bison: Archeology and the First Colonization of Western North America*. Salt Lake City: University of Utah Press, 1999.

———. *Quest for the Origins of the First Americans. Albuquerque*: University of New Mexico Press, 1993.

Dixon, Robert M., and Alexandra Aikhenvald, eds. *Areal Diffusion and Genetic Inheritance: Problems in Comparative Linguistics*. Oxford, UK: Oxford University Press, 2006.

Dongoske, Kurt E. "The Native American Graves Protection and Repatriation Act: A New Beginning, Not the End, for Osteological Analysis—A Hopi Perspective." *Special Issue: Repatriation: An Interdisciplinary Dialogue. American Indian Quarterly* 20, no. 2 (Spring 1996): 287–96.

Dongoske, Kurt E., Michael Yeatts, Roger Anyon, and T. J. Ferguson. "Archaeological Cultures and Cultural Affiliation: Hopi and Zuni Perspectives in the American Southwest." *American Antiquity* 62, no. 4 (1997): 600–608.

Dongoske, Kurt E., and Larry Zimmerman. "Working Together on Race and Racialism in American Archaeology." *SAA Archaeological Record* 10, no. 3 (May 2010): 3–4.

Donnelly, Ignatius. *Atlantis: The Antediluvian World. With an introduction by E. F. Bleiler.* New York: Dover Publications, 1976. First published in 1882 by Harper & Brothers.

Doran, Glen H., David N. Dickel, William E. Ballinger Jr., O. Frank Agee, Phillip J. Laipis, and William W. Hauswirth. "Anatomical, Cellular and Molecular Analysis of 8,000-yr-old Human Brain Tissue from the Windover Archaeological Site." *Nature* 323, no. 6091 (1986): 803–6.

Downey, Roger. *Riddle of the Bones: Politics, Science, Race, and the Story of Kennewick Man.* New York: Copernicus Books, 2000.

Dreger, Alice. *"No Science, Please, We're Anthropologists."* *Psychology Today,* November 25, 2010. http://www.psychologytoday.com/blog/fetishes-i-dont-get/201011/no-science-please-were-anthropologists.

Dumond, D. E. "The Archaeology of Alaska and the Peopling of America." *Science* 209 (1980): 984–91.

———. "A Reexamination of Eskimo-Aleut Prehistory." *American Anthropologist* 89 (1987): 32–56.

Dumont, Clayton W. "The Politics of Scientific Objections to Repatriation." *Wicazo Sa Review* 18, no. 1 (Spring 2003): 109–28.

Egan, Timothy. "Tribe Stops Study of Bones that Challenge History." *New York Times,* September 30, 1996.

Eastern New Mexico University. http://www.enmu.edu/services/museums/blackwater-draw/locality.shtml.

Easton, R. D., D. A. Merriwether, D. E. Crews, and R. E. Ferrell. "mtDNA Variation in the Yanornami: Evidence for Additional New World Founding Lineages." *American Journal of Human Genetics* 59 (1996): 213–25.

Echo-Hawk, Roger. "Ancient History in the New World: Integrating Oral Traditions and the Archaeological Record in Deep Time." *American Antiquity* 65, no. 2 (April 2000): 267–90.

———. "Forging a New Ancient History for Native America." In *Native Americans and Archaeologists: Stepping Stones to Common Ground,* edited by Nina Swidler, Kurt E. Dongoske, Roger Anyon, and Alan S. Downer, 88–102. Walnut Creek, CA: AltaMira, 1997.

———. "Merciless Greetings, Wicked Servants of the Age of Archaeoracialism." *SAA Archaeological Record* 10, no. 3 (May 2010): 21–25.

———. "Working Together on Race." *SAA Archaeological Record,* 10, no. 3 (May 2010): 6–9.

Echo-Hawk, Roger, and Larry J. Zimmerman. "Beyond Racism: Some Opinions about Racialism and American Archaeology." *American Indian Quarterly* 30, no. 3.4 (2006): 461–85.

Edgar, Heather J. H. "Biohistorical Approaches to 'Race' in the United States: Biological Distances among African Americans, European Americans, and Their Ancestors." *American Journal of Physical Anthropology* 139 (2009): 58–67.

Eldredge, Niles, and Steven Jay Gould. "Punctuated Equilibria: An Alternative to Phyletic Gradualism." In *Models in Paleobiology*, edited by Thomas J. M. Schopf, 82–115. San Francisco: Freeman, Cooper and Co., 1972.

Epstein, Steven. "The Construction of Lay Expertise: AIDS Activism and the Forging of Credibility in the Reform of Clinical Trials." *Science, Technology, & Human Values* 20, no. 4 (October 1995): 408–37.

Ericson, Jonathan E., R. Taylor, and Rainer Berger, eds. *Peopling of the New World*. San Diego: Ballena Press, 1982.

Erlandson, Jon M., Torben C. Rick, René L. Vellanoweth, and Douglas J. Kennett. "Maritime Subsistence at a 9300-Year-Old Shell Midden on Santa Rosa Island, California." *Journal of Field Archaeology* 26, no. 3 (Autumn 1999): 255–65.

Fagan, Brian M. *The Great Journey: The Peopling of Ancient America*. New York: Thames and Hudson, 1987.

Fagundes, Nelson J. R., Ricardo Kanitz, Roberta Eckert, Ana C. S. Valls, Mauricio R. Bogo, Francisco M. Salzano, David Glenn Smith, et al. "Mitochondrial Population Genomics Supports a Single Pre-Clovis Origin with a Coastal Route for the Peopling of the Americas." *American Journal of Human Genetics* 82 (2008): 583–92.

Feder, Kenneth L. *Frauds, Myths, and Mysteries: Science and Pseudoscience in Archaeology*. 3rd ed. Mountain View, CA: Mayfield Publishing Co., 1999.

Ferguson, T. J. "Native Americans and the Practice of Archaeology." *Annual Review of Anthropology* 25 (1996): 63–79.

Feyerabend, Paul. *Against Method*. 4th ed. London: Verso, 2010.

Feynman, Richard. "The Development of Space-Time View of Quantum Electrodynamics." Nobel Lecture, December 11, 1965. In *Nobel Lectures: Physics 1963–1970*. 1972, 155.

———. Interview, 1981. In Jeffery Robbins, ed. *The Pleasure of Finding Things Out: The Best Short Works of Richard Feynman, 186–97*. New York: Basic Books, 1999.

Fiedel, Stuart J. "Artifact Provenance at Monte Verde: Inconsistencies and Confusion." *Scientific American Discovering Archaeology* (October 1999).

———. "The Kennewick Follies: 'New' Theories about the Peopling of the Americas." *Journal of Anthropological Research* 60, no. 1 (2004): 75–110.

———. "Older than We Thought: Implications of Corrected Dates for Paleoindians." *American Antiquity* 64 (1999): 95–116.

———. *Prehistory of the Americas*. 2nd ed. Cambridge, UK: Cambridge University Press, 1992.

Fifield, Terence E. "Human Remains Found in Alaska Reported to be 9,730 Years Old." *SAA Bulletin* 14, no. 5 (1996): 5.

Figgins, Jesse D. "The Antiquity of Man in America." *Natural History* 27 (1927): 229–239.

———. "A Further Contribution to the Antiquity of Man in America." *Proceedings of the Colorado Museum of Natural History* 12 (1933): 4–8.

Fisher, Raymond H. *Bering's Voyages: Whither and Why.* Seattle: University of Washington Press, 1977.

Fladmark, Knut R. "Getting One's Berings." *Natural History* 95 (1986): 8–19.

———. "Routes: Alternative Migration Corridors for Early Man in North America." *American Antiquity* 44 (1979): 55–69.

———. "Times and Places: Environmental Correlates of Mid-to-Late Wisconsinan Human Population Expansion in North America." In *Early Man in the New World*, edited by Richard Shutler, 13–42. Beverly Hills, CA: Sage Publications, 1983.

Fletcher, Alice, and Tilly (Matilda) C. Stevenson. "Report of the Committee on the Preservation of Archaeological Remains on the Public Lands." *Proceedings of the American Association for the Advancement of Science* 37 (1889): 35–37.

Flint, Roger Foster. *Glacial and Quaternary Geology.* New York: John Wiley and Sons, 1971.

Forster, P., R. Harding, A. Torroni, and H. J. Bandelt. "Origin and Evolution of Native American mtDNA Variation: A Reappraisal." *American Journal of Human Genetics* 59 (1996): 935–45.

Fortescue, Michael. *Language Relations across Bering Strait.* London: Cassell, 1998.

Foster, John Wells. *Pre-Historic Races of the United States of America.* Chicago: S. C. Griggs and Company, 1873.

Foucault, Michel. *The Order of Things: An Archaeology of the Human Sciences.* New York: Vintage Books, 1970.

———. *Madness and Civilization: A History of Insanity in the Age of Reason.* New York: Vintage, 1988.

Fowler, Don. *A Laboratory for Anthropology: Science and Romanticism in the American Southwest: 1346–1930.* Albuquerque: University of New Mexico Press, 2000.

Fox, C. L. "Mitochondrial DNA Haplogroups in Four Tribes from Tierra del Fuego-Patagonia: Inferences about the Peopling of the Americas." *Human Biology* 68, no. 6 (1996): 855–71.

Fox, J. A. Comment on *The Settlement of the Americas: A Comparison of the Linguistic, Dental, and Genetic Evidence* by Joseph H. Greenberg, Christy G. Turner II, and Stephen L. Zegura. *Current Anthropology* 27 (1986): 477–97.

———. "Man's Antiquity in North America: Views and Facts." *Quaternaria* 9 (1967): 225–35.

Fox, M. *"First Americans May Have Arrived 40,000 Years Ago." Reuters News Service Report*, February 16, 1998.

Freeman, David H. *Wrong: Why Experts Keep Failing Us—and How to Know When Not to Trust Them.* New York: Little Brown, 2010.

Friedman, Janet L., ed. "A History of the Archaeological Resources Protection Act: Laws and Regulations." *American Archeology* 5, no. 2 (1985): 82–119.

Fujisnara, Joan. "Authorizing Knowledge in Science and Anthropology." *American Anthropologist* 100 (1998): 347–60.

Galison, Peter, and David J. Stump, eds. *The Disunity of Science: Boundaries, Contexts, and Power*. Stanford, CA: Stanford University Press, 1996.

Galtung, Johan. "Scientific Colonialism." *Transition* no. 30 (1967): 10–15.

Galvano (Galvao), Antonio. *The Discoveries of the World, from Their First Original unto the Year of Our Lord 1555*, by Antonio Galvano, Governor of Ternate, corrected, quoted, and published in England by Richard Hakluyt, now reprinted with the original Portuguese text (1563), and edited by Vice-Admiral Bethune. Works of the Hakluyt Society, 1st ser., no. 30. New York: B. Franklin, 1601.

George, Debra, and John Southon. "Resolving an Anomalous Radiocarbon Determination on Mastodon Bone from Monte Verde, Chile." *American Antiquity* 70, no. 4 (October 2005): 766–72.

Gibbons, A. "DNA Enters Dust Up Over Bones." *Science* 274 (1996): 172.

Gibson, Charlie. "*EXCERPTS: Charlie Gibson Interviews Sarah Palin*." *ABC World News*, September 11, 2008. http://abcnews.go.com/Politics/Vote2008/story?id=5782924&page=3.

Gilbert, M. Thomas P., Dennis L. Jenkins, Anders Götherstrom, Nuria Naveran, Juan J. Sanchez, Michael Hofreiter, Philip Francis Thomsen, et al. "DNA from Pre-Clovis Human Coprolites in Oregon, North America." *Science* 320 (2008): 786–89.

Giles, R. E., H. Blanc, H. M. Cann, and D. C. Wallace. "Maternal Inheritance of Human Mitochondrial DNA." *Proceedings of the National Academy of Sciences* 77 (1980): 6715–19.

Gill, George W. "Craniofacial Criteria in the Skeletal Attribution of Race." In *Forensic Osteology: Advances in Identification of Human Remains*, edited by Kathleen J. Reichs, 293–317. 2nd ed. Springfield, IL: Charles C. Thomas, 1998.

Gillespie, Susan D., and Deborah L. Nichol, eds. *Archaeology is Anthropology: Archaeological Papers. No. 13*. Arlington, VA: American Anthropological Association, 2003.

Givens, R. Dale. "On the Peopling of America." *Current Anthropology* 9 (1968): 120.

Goebel, Ted, Michael R. Waters, and Dennis H. O'Rourke. "The Late Pleistocene Dispersal of Modern Humans in the Americas." *Science* 319 (2008): 1497–1501.

Goldberg, Paul, and Trina L. Arpin. "Micromorphological Analysis of Sediments from Meadowcroft Rockshelter, Pennsylvania: Implications for Radiocarbon Dating." *Journal of Field Archaeology* 26, no. 3 (1999): 325–42.

Goldstein, Lynne, and Keith Kintigh. "Ethics and the Reburial Controversy." *American Antiquity* 55, no. 3(1990): 585–91.

Golinski, Jan. *Making Natural Knowledge: Constructivism and the History of Science*. Chicago: University of Chicago Press, 2005.

González-José, Rolando, Maria Cátira Bortolini, Fabrício R. Santos, and Sandra L. Bonatto. "The Peopling of America: Craniofacial Shape Variation on a Continental Scale and Its Interpretation from an Interdisciplinary View." *American Journal of Physical Anthropology* 137 (2008): 175–87.

Gosden, Christopher. *Anthropology and Archaeology: A Changing Relationship*. London: Routledge, 1999.

Gould, Stephen Jay. *The Mismeasure of Man*. New York: W. W. Norton & Co., 1981.

———. *Leonardo's Mountain of Clams and the Diet of Worms*. New York: Harmony House, 1998.

Grayson, Donald K. "Perspectives on the Archaeology of the First Americans." In *Americans Before Columbus: Ice Age Origins*, edited by Ronald C. Carlisle, 107–23. Pittsburgh: University of Pittsburgh, 1988.

Grayson Donald K., and David J. Meltzer. "A Requiem for North American Overkill." *Journal of Archaeological Science* 30, no. 5 (2003): 585–593.

Greenberg, Joseph H. *Language in the Americas*. Stanford, CA: Stanford University Press, 1987.

Greenberg, Joseph H., Christy G. Turner II, and Stephen L. Zegura. "The Settlement of the Americas: A Comparison of the Linguistic, Dental, and Genetic Evidence." *Current Anthropology* 27 no. 5 (1986): 477–79.

Griffin, James B. "Comments on the Quest for Early Man in North America." In *Habitats Humains Anterieurs a l'Holocene en Amerique*, edited by James B. Griffin, 7–17. Nice: International Union of Prehistoric and Protohistoric Sciences, 1976.

———. "The Origin and Dispersion of American Indians in North America." In *The First Americans' Origins, Affinities and Adaptations*, edited by William S. Laughlin and Albert B. Harper, 43–55. New York: Gustav Fischer, 1979.

Gruhn, Ruth. "Language Classification and Models of the Peopling of the Americas." In *Archaeology and Linguistics: Aboriginal Australia in Global Perspective*, edited by Patrick McConvell and Nicholas Evans, 99–110. Melbourne: Oxford University Press, 1997.

———. "Linguistic Evidence in Support of the Coastal Route of the Earliest Entry into the New World." *Man* 23 (1988): 77–100.

———. "On the Settlement of the Americas: South American Evidence for an Expanded Time Frame." *Current Anthropology* 28 (1987): 363–64.

———. "The Pacific Coast of Initial Entry: An Overview." In *Method and Theory for Investigating the Peopling of the Americas*, edited by Robson Bonnichsen and D. Gentry Steele, 249–56. Corvallis: Center for the Study of the First Americans, Oregon State University, 1994.

Guidon, N., and G. Delibrias. "Carbon-14 Dates Point to Man in the Americas 32,000 Years Ago." *Nature* 321 (1986): 769–71.

———. "Faunal Exchanges between Siberia and North America: Evidence from Quaternary Land Mammal Remains in Siberia, Alaska and the Yukon Territory." *Canadian Journal of Anthropology* 1 (1980): 45–49.

Gullapalli, Praveena. Review of *Who Owns Antiquity? Museums and the Battle over Our Ancient Heritage* by James Cuno. *Expedition* 51, no. 2 (2009): 48.

Habu, Junk, Clare Fawcett, and John M. Matsunaga, eds. *Evaluating Multiple Narratives: Beyond Nationalist, Colonist, Imperialist Archaeologies*. New York: Springer, 2008.

Hacking, Ian. *Representing and Intervening: Introductory Topics in the Philosophy of Natural Science*. Cambridge, UK: Cambridge University Press, 1983.

Hakluyt, Richard. *Divers Voyages Touching on the Discovery of America*. London: Thomas Woodcocke, 1582.

Hall, D. "Corrected Radiocarbon Calendar Can Clarify Peopling of Americas." *Mammoth Trumpet* 12, no. 4 (1997): 4–7.

Haraway, Donna. *Simians, Cyborgs and Women: The Reinvention of Nature*. London: Free Association Books, 1991.

Harding, Sandra. *Is Science Multi-Cultural?* Bloomington: Indiana University Press, 1998.

Harmon, Amy. "Where'd You Go with My DNA?" *New York Times*, April 24, 2010. http://www.nytimes.com/2010/04/25/weekinreview/25harmon.html.

Harper, A., and W. Laughlin. "Inquiries into the Peopling of the New World: Development of Ideas and Recent Advances." In *A History of American Physical Anthropology*, edited by Frank Spencer, 281–304. New York: Academic Press, 1982.

Hatab, Lawrence J. *Myth and Philosophy: A Contest of Truths*. La Salle, IL: Open Court Publishing, 1990.

Haven, Samuel Foster. *The Archaeology of the United States, or Sketches, Historical and Bibliographical, of the Progress of Information and Opinion Respecting Vestiges of Antiquity in the United States. Smithsonian Contributions to Knowledge*. Washington, DC: Smithsonian Institution, 1856.

Hawking, Stephen. *A Brief History of Time: From the Big Bang to Black Holes*. New York: Bantam Books, 1988.

Haynes, C. V., Jr. "Carbon-14 Dates and Early Man in the New World." In *Pleistocene Extinctions*, edited by P. S. Martin and H. E. Wright, 267–86. New Haven: Yale University Press, 1967.

———. "The Clovis Culture." *Canadian Journal of Anthropology* 1 (1980): 115–21.

———. "Clovis Origin Update." *The Kiva* 52 (1987): 83–93.

———. Comment on *Early Man in America and the Late Pleistocene Chronology of Western Canada and Alaska* by A. L. Bryan. *Current Anthropology* 10 (1969): 353–54.

———. "Contributions of Radiocarbon Dating to the Geochronology of the Peopling of the New World." In *Radiocarbon after Four Decades*, edited by Renee Kra, 355–74. New York: Springer-Verlag, 1992.

———. "The Earliest Americans." *Science* 166 (1969): 709–15.

———. "Fluted Projectile Points; Their Age and Dispersion." *Science* 145 (1964): 1408–13.

———. "Geoarchaeological and Paleohydrological Evidence for a Clovis-Age Drought in North America and Its Bearing on Extinction." *Quaternary Research* 35 (1991): 438–50.

———. "When and From Where Did Man Arrive in Northeastern North America: A Discussion." In *Amerinds and Their Paleoenvironments in Northeastern North America*, edited by Walter S. Newman and Bert Salwen. *Annals of the New York Academy of Sciences* 288: 165–66. New York: New York Academy of Sciences, 1977.

Haynes, C. V., Jr., and George A. Agogino. *Geochronology of Sandia Cave*. Washington, DC: Smithsonian Institution Press, 1986.

Haynes, Gary A. *The Early Settlement of North America: The Clovis Era*. Cambridge, UK: Cambridge University Press, 2002.

Heckewelder, John. *History, Manners, and Customs of the Indian Nations Who Once Inhabited Pennsylvania and the Neighboring States*. Rev. ed., with an introduction and notes by the Rev. William C. Reichel. Philadelphia: Historical Society of Pennsylvania, 1819.

Heinz History Center. Accessed June 27, 2019. www.heinzhistorycenter.org/about.

Hester, J. J. *Blackwater, Locality No. 1*. Dallas: Southern Methodist University, 1972.

Hinsley, Curtis M., Jr. *Savages and Scientists: The Smithsonian Institution and the Development of American Anthropology, 1846–1910*. Washington, DC: Smithsonian Institution Press, 1981.

———. *The Smithsonian and the American Indian: Making a Moral Anthropology in Victorian America*. Washington, DC: Smithsonian Institution Press, 1994.

Hobbes, Thomas. "Of the Natural Condition of Mankind as Concerning Their Felicity and Misery." In *Leviathan or The Matter, Forme and Power of a Common Wealth Ecclesiasticall and Civil*. Cambridge, UK: Cambridge University Press, 1651. http://oregonstate.edu/instruct/phl302/texts/hobbes/leviathan-c.html.

Hodder, Ian, ed. *The Meanings of Things: Material Culture and Symbolic Expression*. London: Harper Collins, 1989.

Hoffecker, John F., and Scott A. Elias. *Human Ecology of Beringia*. New York: Columbia University Press, 2007.

Hoeffecker, John F., W. Roger Powers, and Ted Goebel. "The Colonization of Beringia and the Peopling of the New World." *Science* 259 (1993): 46–53.

Holmes, W. H. "Traces of Glacial Man in Ohio." *Journal of Geology* 1 (1893): 147–163.

———. *Handbook of Aboriginal American Antiquities*. Part 1, Introductory: The Lithic Industries. Bulletin of the Bureau of American Ethnology 60. Washington, DC: US Government Printing Office, 1919.

———. "The Antiquity Phantom in American Archaeology." *Science* 62 (1925): 256–258.

Horn, Amanda. "The Kennewick Man Loses Sleep over NAGPRA: Native Americans and Scientists Wrestle over Cultural Remains." In *Sovereignty Symposium X*. Oklahoma City: Oklahoma Bar Association, 1997.

Howells, William White. *Who's Who in Skulls: Ethnic Identification of Crania from Measurements*. Cambridge, MA: Harvard University, 1995.

Hrdlička, Aleš. "The Crania of Trenton, New Jersey and Their Bearing upon the Antiquity of Man in That Region." *American Museum of Natural History Bulletin* 16 (1902): 23–62.

———. "Early Man in America: What Have the Bones to Say?" In *Early Man as Depicted by Leading Authorities at the International Symposium at the Academy of Natural Sciences*, Philadelphia, March 1937, edited by G. G. MacCurdy, 93–104. Philadelphia: J. B. Lippincott, 1937.

———. *Early Man in South America.* In collaboration with William Holmes. Washington, DC: US Government Printing Office, 1912.

———. "The Origin and Antiquity of the American Indian." In *Annual Report of the Board of Regents of the Smithsonian Institution for 1923,* 481–93. Washington, DC: US Government Printing Office, 1925.

———. "Preliminary Report on Finds of Supposedly Ancient Human Remains at Vero, Florida." *Journal of Geology* 25 (1917): 43–51.

———. "The Problem of Man's Antiquity in America." *Proceedings of the Eighth American Scientific Congress held in Washington May 10–18, 1940, under the Auspices of the Government of the United States of America . . . ,* edited by Paul H. Oehser, 53–55. Washington, DC: Department of State, 1941–43.

———. "The Race and Antiquity of the American Indian." *Scientific American* 135 (1926): 7–9.

———. "Recent Discoveries Attributed to Early Man in America." *Bureau of American Ethnology Bulletin* 66 (1918): 1–67.

———. "Skeletal Remains Suggesting or Attributed to Early Man in North America." *Bureau of American Ethnology Bulletin* 33 (1907): 1–113.

———. "Studies on Early Man in Europe." In *Smithsonian Explorations, 1923. Smithsonian Miscellaneous Collections* 76, no. 10 (1925): 56–66.

Huckell, Bruce B., and J. David Kilby. "Late Pleistocene North America and Early Paleoindians." In *Readings in Late Pleistocene North America and Early Paleoindians: Selections from American Antiquity,* edited by Bruce B. Huckell and J. David Kilby, viii. Washington, DC: Society for American Archaeology, 2004.

Huckell, Bruce B., and J. David Kilby, eds. *Readings in Late Pleistocene North America and Early Paleoindians: Selections from American Antiquity.* Washington, DC: Society for American Archaeology, 2004.

Huddleston, Lee Eldridge. *Origins of the American Indians: European Concepts, 1492–1729. Austin:* University of Texas Press, 1967.

Humboldt, Alexander Von. *Researches Concerning the Institutions and Monuments of the Ancient Inhabitants of America with Descriptions and Views of Some of the Most Striking Scenes in the Cordilleras! 2 vols. Translated by Helen Maria Williams.* London: Longman, Hurst, Rees, Orme & Brown, J. Murray & H. Colburn, 1814.

Ioannidis, John P. A. "Why Most Published Research Findings Are False." *PLoS Medicine* 2, no. 8 (2005): 696–701.

Iowa County Historical Society. https://iowacountyhistoricalsociety.org/.

Irving, W. N. "Context and Chronology of Early Man in the Americas." In *Annual Review of Anthropology,* vol. 14, edited by B. J. Siegel, A. R. Beals, and S. A. Tyler, 529–55. Palo Alto, CA: Annual Reviews, 1985.

———. "New Dates from Old Bones." *Natural History* 96, no. 2 (1987): 8–13.

Jameson, John H., Jr., ed. *The Reconstructed Past: Reconstructions in the Public Interpretation of Archaeology and History.* Walnut Creek, CA: Alta Mira, 2004.

Jantz, R. L., D. R. Hunt, A. B. Falsetti, and P. J. Key. "Variation among North Amerindians: Analysis of Boas's Anthropometric Data." *Human Biology* 64 (1992): 435–61.

Jantz, R. L., and Douglas W. Owsley. "Pathology, Taphonomy, and Cranial Morphometrics of the Spirit Cave Mummy." *Nevada Historical Quarterly* 40 (1997): 62–84.

———. "Variation among Early North American Crania." *American Journal of Physical Anthropology* 114 (2001): 146–55.

Jefferson, Thomas. *Notes on the State of Virginia. 1787. Edited with an introduction and notes by William Peden. Chapel Hill*: University of North Carolina Press, 1954.

Jelinek, Arthur J. "Perspectives from the Old World on the Habitation of the New." In *Readings in Late Pleistocene North America and Early Paleoindians: Selections from American Antiquity*, edited by Bruce B. Huckell and J. David Kilby, 45.

Johnson, George. "Indian Tribes' Creationists Thwart Archeologists." *New York Times*, October 22, 1996.

Johnson, Greg. "Naturally There: Discourses of Permanence in the Repatriation Context." *History of Religions* 44, no. 1 (August 2004): 36–55.

Johnston, W. A. "Quaternary Geology of North America in Relation to the Migration of Man." In *The American Aborigines: Their Origin and Antiquity*, edited by Diamond Jenness, 9–45. Toronto: University of Toronto Press, 1933.

Jones, Peter, and Darby C. Stapp. "An Anthropological Perspective on Magistrate Jelderks' Kennewick Man Decision." *High Plains Applied Anthropologist* 23, no. 1 (2004): 1–16.

Jones, Rhys. "From Kakadu to Kutikina: The Southern Continent at 18,000 Years Ago." In *The World at 18,000 BP: Low Latitudes*, edited by Olga Soifer and Clive Gamble, 264–295. London: Unwin Hyman, 1990.

Kakaliouras, Ann M. "Race Is . . . Only as Race Does: Essentialism and Ethnicity in (Bio)Archaeology and Skeletal Biology." *SAA Archaeological Record* 10, no. 3 (May 2010): 16–20.

———. "Leaving Few Bones Unturned: Recent Work on Repatriation by Bioarchaeologists." *American Anthropologist* 110, no. 1 (2008): 44–52.

———. "Toward a 'New and Different' Osteology: A Reflexive Critique of Physical Anthropology in the United States since the Passage of NAGPRA." In *Opening Archaeology: Repatriation's Impact on Contemporary Research and Practice*, edited by Thomas W. Killion, 119. Santa Fe, NM: School for Advanced Research Press, 2008.

Katz, Lienke. *The History of Blackwater Draw. Portales: Eastern New Mexico University Printing Services*, 1997.

Kemp, Brian M., Ripan S. Malhi, John McDonough, Deborah A. Bolnick, Jason A. Eshleman, Olga Rickards, Cristina Martines-Labarga, et al. "Genetic Analysis of Early Holocene Skeletal Remains from Alaska and Its Implications for the Settlement of the Americas." *American Journal of Physical Anthropology* 132 (2007): 605–21.

Kennicott, Philip. "*Following Controversy, Mummies at Penn Museum Remain Objects of Mystery.*" *Washington Post*, March 3, 2011. www.washingtonpost.com/wp-dyn/content/article/2011/03/02/AR2011030206371.html.

Kidder, A. V. "Early Man in America." *The Masterkey 1*, no. 5 (1927): 5–13.

Kilby, David. *Eastern New Mexico University Field Projects*. http://academic.enmu. edu/kilbyd/fieldprojects.html.

Killick, David, and Paul Goldberg. "A Quiet Crisis in American Archaeology." *SAA Archaeological Record* 9, no. 1, (January 2009): 6–10, 40.

Killion, Thomas W., ed. *Opening Archaeology: Repatriation's Impact on Contemporary Research and Practice*. Santa Fe, NM: School for Advanced Research Press, 2008.

King, Cecil. "Here Come the Anthros." In *Indians and Anthropologists: Vine Deloria, Jr., and the Critique of Anthropology*, edited by Thomas Biolsi and Larry J. Zimmerman, 115–19. Tucson: University of Arizona Press, 1997.

King, Thomas F. "Resolving a Conflict of Values in American Archaeology." In *Conservation Archaeology*, edited by Michael B. Schiffer and George J. Gumerman, 87–95. New York: Academic, 1977.

———. *Cultural Resource Laws and Practice: An Introductory Guide*. 2nd ed. Walnut Creek, CA: AltaMira, 2004.

Kintigh, Keith W. "Repatriation as a Force of Change in Southwestern Archaeology." In *Opening Archaeology: Repatriation's Impact on Contemporary Research and Practice*, edited by Thomas W. Killion, 204. Santa Fe, NM: School for Advanced Research Press, 2008.

Kitchen A., M. M. Mlyamoto, and M. I. Mulligan. "A Three-Step Colonization Model for the Peopling of the Americas." *PLoS ONE* 3 (2008): 1–7.

Knorr-Cetina, Karin D. *The Manufacture of Knowledge: An Essay on the Constructivist and Contextual Nature of Science*. New York: Pergamon Press, 1981.

Knudson, Kelly J., and Christopher M. Stojanowski, eds. *Bioarchaeology and Identity in the Americas*. Gainesville: University Press of Florida, 2009.

Knudson, Ruthann, and Bennie C. Keel, eds. *The Public Trust and the First Americans*. Corvallis: Center for the Study of the First Americans, Oregon State University, 1981.

Koerner, Stephanie. "Archaeology, Nationalism, and Problems Posed by Science/ Values, Epistemology/Ontology Dichotomies." *World Archaeology Bulletin* (August–September 2001): 57–96.

Koch, David. Accessed February 15, 2011. http://www.kochind.com/files/KochDavid.pdf.

Kolman, C. J., N. Sambuughin, and E. Bermingham. "Mitochondrial DNA Analysis of Mongolian Populations and Implications for the Origin of New World Founders." *Genetics* 142 (1996): 1321–34.

Kuhn, Thomas S. *The Structure of Scientific Revolutions*. Chicago: University of Chicago Press, 1962.

Lahr, Marta Mirazon. "Patterns of Modern Human Diversification: Implications for Amerindian Origins." *Yearbook of Physical Anthropology* 38 (1995): 163–98.

Lambert, Joseph B. *Traces of the Past: Unraveling the Secrets of Archaeology through Chemistry*. Cambridge, MA: Perseus Press, 1997.

Landau, Patricia M., and D. Gentry Steele. "Why Anthropologists Study Human Remains." In *Repatriation Reader: Who Owns American Indian Remains?* Edited by Devon A. Mihesuah, 74–94. Lincoln: University of Nebraska Press, 2000.

Larsen, Clark S. *Bioarchaeology: Interpreting Behavior from the Human Skeleton.* Cambridge, UK: Cambridge University Press, 1997.

Latour, Bruno. "Give Me a Laboratory and I Will Raise the World." In *Science Observed: Perspectives on the Social Studies of Science,* edited by Karin D. Knorr-Cetina and Michael Mulkay. London: Sage, 1983.

Latour, Bruno, and Steve Woolgar. *Laboratory Life: The Construction of Scientific Facts.* Princeton, NJ: Princeton University Press, 1986.

Laudan, Larry. *Progress and Its Problems: Towards a Theory of Scientific Growth.* Berkeley: University of California Press, 1977.

Laughlin, W. S. Comment on *The Settlement of the Americas: A Comparison of the Linguistic, Dental, and Genetic Evidence* by Joseph H. Greenberg, Christy G. Turner II, and Stephen L. Zegitra. *Current Anthropology* 27 (1986): 489–90.

Lavallee, D. *The First South Americans: The Peopling of a Continent from the Earliest Evidence to High Culture.* Translated by P. G. Bahn. Salt Lake City: University of Utah Press, 2000.

Lawson, John. *A New Voyage to Carolina: Containing the Exact Description of That Country: Together with the Present State Thereof, and a Journal of a Thousand Miles, Travl'd thro' Several Nations of Indians, Giving a Particular Account of Their Customs, Manners, &c.* London: s.n., 1709.

Leclerc, Georges-Louis, Comte de Buffon. *Histoire naturelle, générale et particulière (1749–1789).*

Lee, R. F. *The Antiquities Act.* Washington, DC: *National Park Service, Department of the Interior,* 1970.

Lell J. T., M. D. Brown, T. G. Schurr, R. I. Sukernik, E. B. Starikovskaya, A. Torroni, and D. C. Wallace. "Y-Chromosome Polymorphisms in Native American and Siberian Populations: Identification of Founding Native American Y-Chromosome Haplotypes." *Human Genetics* 100 (1997): 536–43.

Leone, Mark P., and Parker B. Potter. "Legitimation and the Classification of Archaeological Sites." *American Antiquity* 57 (1992): 137–45.

Lepper, Bradley, and Robson Bonnichsen, eds. *New Perspectives on the First Americans.* College Station: Texas A&M University Press, 2004.

Lewin, Roger. "Skepticism Fades over Pre-Clovis Man." *Science* 244, no. 4909 (June 9, 1989): 1140.

Lewis, C. M. "Mitochondrial DNA and the Peopling of South America." *Human Biology* 79 (2007): 159–79.

Lewis, Jason E., David DeGusta, Marc R. Meyer, Janet M. Monge, Alan E. Mann, and Ralph L. Holloway. "The Mismeasure of Science: Stephen Jay Gould versus Samuel George Morton on Skulls and Bias." *PLoS Biology* 9, no. 6 (June 2011): 1–6.

Li, Chunxiang, Li Hongjie, Yinqiu Cui, Chengzhi Xie, Dawei Cai, Wenying Li, Victor H. Mair, et al. "Evidence that a West-East Admixed Population Lived in the Tarim Basin as Early as the Early Bronze Age." *BMC Biology* 8 (2010): 15.

Lloyd, Geoffrey E. R. *Demystifying Mentalities.* Cambridge, UK: Cambridge University, 1990.

Long, John. *Voyages and Travels of an Indian Interpreter and Trader, Describing the Manners and Customs of the North American Indians, with an Account of the Posts Situated on the River Saint Lawrence, Lake Ontario, etc. to Which Is Added, a Vocabulary of the Chippeway Language, Names of Furs and Skins, in English and French, a List of Words in the Iroquois, Mohegan, Shawanee, and Esquimaux Tongues, and a Table Showing the Analogy Between the Algonkin and Chippeway Languages.* London: Printed by Robson and others, 1791.

Longino, Helen E. *Science as Social Knowledge: Values and Objectivity in Scientific Inquiry.* Princeton, NJ: Princeton University Press, 1990.

Lorenz, Joseph G., and David Glenn Smith. "Distribution of Four Founding mtDNA Haplogroups among Native North Americans." *American Journal of Physical Anthropology* 101 (1996): 307–23.

Lyell, Charles. *The Geological Evidence of the Antiquity of Man.* London: J. M. Dent & Sons, 1914. First published in 1863 by John Murray.

Lynch, Thomas F. "Glacial-Age Man in South America? A Critical Review." *American Antiquity* 55 (1990): 12–36.

———. "Lack of Evidence for Glacial Age Settlement of South America: Reply to Dillehay and Collins and to Gruhn and Bryan." *American Antiquity* 56, no. 2 (April 1991): 348–55.

MacCurdy, George Grant. *Early Man.* Philadelphia: Lippincott, 1937.

———. "The Problem of Man's Antiquity at Vero, Florida." *American Anthropologist* 19 (1916): 252–61.

MacGowan, Kenneth, and Joseph A. Hester Jr. *Early Man in the New World.* Garden City, NY: Doubleday and Company, 1962.

Macneiei-I, Rinnaan S. "Men Out of Asia: As Seen from the Northwest Yukon." *Anthropological Papers of the University of Alaska* 7, no. 2 (1959): 41–70.

MacNeish, R. "Early Man in the New World." *American Scientist* 63 (1976): 316–27.

Madsen, David B., ed. *Entering America: Northeast Asia and Beringia Before the Last Glacial Maximum.* Salt Lake City: University of Utah Press, 2004.

Mahli, R. S., Brian M. Kemp, Jason A. Eshleman, Jerome Cybulski, David Glenn Smith, Scott Cousins, and Harold Harry. "Mitochondrial Haplogroup M Discovered in Prehistoric North Americans." *Journal of Archaeological Science* 34, no. 4 (2007): 642–48.

Mair, Victor H. "Ancient Mummies of the Tarim Basin" *Expedition Magazine* 58, no. 2 (2016). http://www.penn.museum/sites/expedition/?p=23790.

Mandelbaum, Maurice. "Causal Analysis in History." *Journal of the History of Ideas* 3, no. 1 (January 1942): 30–50.

Mandryk, Carole. *"Paleoecology as Contextual Archaeology: Human Viability of the Late Quaternary Ice-Free Corrido, Alberta, Canada."* PhD diss., University of Alberta, 1992.

Mandryk, Carole A. S., Heiner Josenhans, Daryl W. Fedje, and Rolf W. Mathewes. "Late Quaternary Paleoenvironments of Northwestern North America: Implications for Inland versus Coastal Migration Routes." *Quaternary Science Reviews* 20 (2001): 301–314.

Marshall, Eliot. "Clovis Counterrevolution." *Science* 249, no. 4970 (August 17, 1990): 738–41.

———. "Pre-Clovis Sites Fight for Acceptance." *Science* 291, no. 5590 (March 2, 2001): 1730–32.

———. "Whose DNA Is It, Anyway?" *Science* 278 (1997): 564–67.

Martin, Paul S. "Clovis the Beautiful." *Natural History* 96, no. 10 (1987): 10–13.

———. "The Discovery of America." *Science* 179 (1973): 969–74.

———. *The Last 10,000 Years: A Fossil Pollen Record of the American Southwest*. Tucson: University of Arizona Press, 1963.

———. "Pleistocene Overkill." *Natural History* 76, no. 10 (1967): 32–38.

———. "Prehistoric Overkill." In *Pleistocene Extinctions: The Search for a Cause*, edited by Paul S. Martin and Herb E. Wright Jr., 75–120. New Haven, CT: Yale University Press, 1967.

———. "Prehistoric Overkill: The Global Model." In *Quaternary Extinctions: A Prehistoric Revolution*, edited by Paul S. Martin and Richard G. Klein, 354–403. Tucson: University of Arizona Press, 1984.

———.. *"Who or What Destroyed Our Mammoths?"* In *Megafauna and Man: Discovery of America's Heartland*, edited by Larry D. Agenbroad, 109–17. Malden, MA: Wiley Interscience, 1990.

Martin, Paul S., and Richard G. Klein, eds. *Quaternary Extinctions: A Prehistoric Revolution*. Tucson: University of Arizona Press, 1984.

Mason, Ronald J. *Inconstant Companions: Archaeology and North American Oral Traditions*. *Tuscaloosa*: University of Alabama Press, 2006.

McBryde, Isabel, ed. *Who Owns the Past?* Oxford, UK: Oxford University Press, 1985.

McGee, William J. "Man and the Glacial Period." *American Anthropologist* 6, no. 1 (1893): 85–95.

———. "Paleolithic Man in America: His Antiquity and Environment." *Popular Science Monthly* 34 (1889): 20–57.

McGuire, Randall H. "Archaeology and the First Americans." *American Anthropologist* 94, no. 4 (1992): 816–36.

———. *Archaeology as Political Action*. Berkeley: University of California Press, 2008.

McKeown, C. T. "Principles of Consultation." *CRM* 2, no. 3/4 (1997): 26–27.

McManamon, Francis P. "The Antiquities Act: Setting Basic Preservation Policies." *CRM* 19, no. 7 (1996): 18–23.

———. "Changing Relationships between Native Americans and Archaeologists." *Historic Preservation Forum* 8, no. 2 (1994): 15–20.

———. "Managing Repatriation: Implementing the Native American Graves Protection and Repatriation Act." *CRM* 15, no. 5 (1992): 9–12.

———. "The Many Publics for Archaeology." *American Antiquity* 56, no. 1 (1991): 121–30.

———. "The Native American Graves Protection and Repatriation Act and First Americans Research." In *Who Were the First Americans: Proceedings of the 58th*

*Annual Biology Colloquium*, edited by Ruth Gruhn and Robson Bonnichsen, 145. Corvallis: Center for the Study of the First Americans, Oregon State University, 1999.

———. "*Striking a Balance with NAGPRA.*" *Anthropology Newsletter April*, 1998, 20.

McManamon, F. P., and L. V. Nordby. "Implementing the Native American Graves Protection and Repatriation Act." *Arizona State Law Journal* 24, no. 1 (1992): 217–52.

Mead, Jim I., and David J. Meltzer. *Environments and Extinction: Man in Late Glacial America. Orono: Center for the Study of Early Man, University of Maine*, 1985.

Meltzer, David J. "The Antiquity of Man and the Development of American Archaeology." In *Advances in Archaeological Method and Theory*, vol. 6, edited by Michael B. Schiffer, 1–51. New York: Academic Press, 1983.

———. "Clocking the First Americans." *Annual Review of Anthropology* 24 (1995): 21–45.

———. *First Peoples in a New World: Colonizing Ice Age America*. Berkeley: University of California Press, 2009.

———. "Late Pleistocene Human Adaptations in Eastern North America." *Journal of World Prehistory* 2 (1988): 1–52.

———. "Monte Verde and the Pleistocene Peopling of the Americas." *Science* 276 (1997): 754–55.

———. "North American Archaeology and Archaeologists, 1879–1934." *American Antiquity* 50, no. 21 (1985): 249–60.

———. "Peopling of North America." In *The Quaternary Period in the United States*, edited by Alan R. Gillespie, Stephen C. Porter, and Brian Atwater, 539–63. New York: Elsevier Science, 2004.

———. *Search for the First Americans. Washington, DC: Smithsonian Books*, 1993.

———. "The Seventy-Year Itch: Controversies over Human Antiquity and Their Resolution." *Journal of Anthropological Research* 61, no. 4 (Winter 2005): 433–68.

———. "Weber and Anthropology." *Annual Review of Anthropology* 24 (1995): 21–45.

———. "*Why Don't We Know When the First People Came to North America?*" In *Readings in Late Pleistocene North America and Early Paleoindians: Selections from American Antiquity*, edited by Bruce B. Huckell and J. David Kilby, 26. Washington, DC: Society for American Archaeology, 2004.

Meltzer, David J., and Bruce D. Smith. "Paleo-Indian and Early Archaic Subsistence Strategies in Eastern North America." In *Foraging, Collecting and Harvesting: Archaic Period Subsistence and Settlement in the Eastern Woodlands*, edited by S. Neusius, 1–30. Carbondale: Center for Archaeological Investigations, Southern Illinois University, 1986.

———. "The Pleistocene Archaeology of Beringia." In *The Evolution of Human Hunting*, edited by M. Nitecki and D. Nitecki, 267–307. New York: Plenum, 1987.

Meltzer, David J., and Jim I. Mead. "Dating Late Pleistocene Extinctions: Theoretical Issues, Analytical Bias, and Substantive Results." In *Environments and Extinctions:*

*Man in Late Glacial North America*, edited by Jim I. Mead and David J. Meltzer, 145–73. Orono: Center for the Study of Early Man, University of Maine, 1985.

Meltzer, David J., and William C. Sturtevant. "The Holly Oak Shell Game: An Historic Archaeological Fraud." In *Lulu Linear Punctated: Essays in Honor of George Irving Quimby*, edited by Robert C. Dunnell and Donald K. Grayson, 325–52. Ann Arbor: Museum of Anthropology, University of Michigan, 1983.

Meltzer, David J., Don D. Fowler, and Jeremy A. Sabloff, eds. *American Archaeology Past and Future*. Washington, DC: Smithsonian Institution Press, 1986.

Meltzer, David J., Donald K. Grayson, Gerardo Ardila, Alex W. Barker, Dena F. Dincauze, C. Vance Haynes, Francisco Mena, Lautaro Nuñez, and Dennis J. Stanford. "On the Pleistocene Antiquity of Monte Verde, Southern Chile." *American Antiquity* 62, no. 4 (1997): 659–63.

Meltzer, David J., James M. Adovasio, and Tom D. Dillehay. "On a Pleistocene Human Occupation at Pedra Furada, Brazil." *Antiquity* 68, no. 261 (1994): 695–714.

Merriwether D. A. "Mitochondrial DNA." In *Handbook of North American Indians*, vol. 3, edited by Douglas H. Ubelaker, William C. Sturtevant, and Dennis Stanford, 817–830. Washington, DC: Smithsonian Institute Press, 2007.

Merriwether, D. A., and R. E. Ferrell. "The Four Founding Lineage Hypothesis for the New World: A Critical Reevaluation." *Molecular and Phylogenetic Evolution* 5 (1996): 241–46.

Merriwether, D. A., F. Rothhammer, and R. E. Ferrell. "Distribution of the Four-Founding Lineage Haplotypes in Native Americans Suggests a Single Wave of Migration for the New World." *American Journal of Physical Anthropology* 98 (1995): 411–30.

———. "Genetic Variation in the New World: Ancient Teeth, Bone, and Tissue as Sources of DNA." *Experientia* 50 (1994): 592–601.

Merriwether, D. A., W. W. Hall, A. Vahlne, and R. E. Ferrell. "mtDNA Variation Indicates Mongolia May Have Been the Source for the Founding Populations for the New World." *American Journal of Human Genetics* 59 (1996): 204–12.

Merton, Robert K. *The Sociology of Science: Theoretical and Empirical Investigations*. Chicago: University of Chicago Press, 1973.

Meskell, Lynn. "The Intersections of Identity and Politics in Archaeology." *Annual Reviews in Anthropology* 31 (2002): 279–301.

Mihesuah, Devon, ed. *Repatriation Reader: Who Owns American Indian Remains?* New York: Bison, 2000.

Millet, G., F. Rothhammer, F. R. Carnese, C. M. Bravi, and N. O. Bianchi. "Founder Mitochondrial Haplotypes in Amerindian Populations." *American Journal of Human Genetics* 54 (1994): 27–33.

"'Million-Dollar' Man-Ape's Tooth Shattered; Museum's Prize Falls from X-Ray Man's Hand." *New York Times*, February 19, 1925.

Monsalve, M. V., H. G. de Restrepo, A. E. Spinel, G. Correal, and D. V. Devine. "Evidence of Mitochondrial DNA Diversity in South American Aboriginals." *Annals of Human Genetics* 58 (1994): 265–73.

*"Monte Verde Under Fire."* *Archaeology*, October 18, 1999. http://www.archaeology. org/online/features/clovis/.

Moorehead, Warren King. *The Stone Age in North America. 2 vols.* Boston: Houghton Mifflin, 1910.

Morgan, Kathryn A. *Myth and Philosophy from the Presocratics to Plato.* Cambridge, UK: Cambridge University Press, 2000.

Morell, V. "Genes May Link Ancient Eurasians, Native Americans." *Science* 280 (1998): 520.

———. "Who Owns the Past?" *Science* 268 (1995): 1424–26.

Morlot, Adolphe von. "General Views on Archaeology." In *Annual Report of the Board of Regents of the Smithsonian Institution for 1860*, 284–343. Washington, DC: US Government Printing Office, 1861.

Morrow, Juliet E., and Cristobal Gnecco, eds. *Paleoindian Archaeology: A Hemispheric Perspective.* Gainesville: University Press of Florida, 2006.

Morrow, Juliet E., and Toby A. Morrow. "Geographic Variation in Fluted Projectile Points: A Hemispheric Perspective." *American Antiquity* 64 (1999): 215–30.

Morton, Samuel George. *Crania Americana; or, A Comparative View of the Skulls of Various Aboriginal Nations of North and South America. Philadelphia: J. Dobson*, 1839.

Mosimann, J. E., and P. S. Martin. "Simulating Pleistocene Overkill by Palaeo-Indians." *American Scientist* 63, no. 3 (1975): 304–13.

Mullins, Paul R. "Archaeologies of Invisibility and Neo-Racism." *SAA Archaeological Record* 10, no. 3 (May 2010): 10–11.

Mulvaney, John. *The Prehistory of Australia.* Harmondsworth: Penguin Books, 1975.

Munford, D., M. C. Zanini, and W. W. Neves. "Human Cranial Variation in South America: Implications for the Settlement of the New World." *Brazilian Journal of Genetics* 18 (1995): 673–88.

Naddaf, Gerard. "Allegory and the Origins of Philosophy." In *Logos and Muthos: Philosophical Essays in Greek Literature*, edited by William Wians, 99–131. Albany: State University of New York Press, 2009.

National Park Service Archaeology Program. *Kennewick Website.* Accessed January 26, 2021. http://www.nps.gov/archeology/kennewick/index.htm.

Native American Graves Protection and Repatriation Act, as amended. *25 U.S.C.* 3005. *Nature*, August 26, 1922, 282.

Neiburger, E. J. *"Ethnic Minority Council of America. July 15, 2005, letter to Senator John McCain, Chairman, Senate Committee on Indian Affairs."* Accessed April 12, 2011. http://www.friendsofpast.org/nagpra/news/EMCLetter.pdf.

Neimark, Jill. *"Pinkie Pokes Holes in Human Evolution."* *Discover*, January 2, 2011, 55.

Nelson, D. E., Richard E. Morlan, J. S. Vogel, J. R. Southon, and C. R. Harington. "New Dates on Northern Yukon Artifacts: Holocene not Upper Pleistocene." *Science* 232 (1986): 749–51.

Nelson, N. C. "The Antiquity of Man in America in the Light of Archaeology." In *The American Aborigines: Their Origin and Antiquity*, edited by Diamond Jenness, 87–130. Toronto: University of Toronto Press, 1933.

———. Discussion of *The Origin and Antiquity of Man in America* by Ales Hrdlička. *Bulletin of the New York Academy of Medicine* 4 (1928): 820–23.

Neves, W. A. "Cranial Morphological Variation and the Colonization of the New World: Towards a Four Migration Model." *American Journal of Physical Anthropology* S22 (1996): 176.

Neves, W. A., and H. M. Pucciarelli. "The Zhoukoudian Upper Cave Skull 101 as Seen from the Americas." *Journal of Human Evolution* 34 (1998): 219–22.

Neves, W. A., H. M. Pucciarelli, and D. Meyer. "The Contribution of the Morphology of Early South and North American Skeletal Remains to the Understanding of the Peopling of the Americas." *American Journal of Physical Anthropology*, Supplement 16 (1993): 150–51.

Nicholas, George, and Thomas Andrews, eds. *At a Crossroads: Archaeology and First Peoples in Canada*. Burnaby, BC: Archaeology Press, 1997.

Nichols, J. *Linguistic Diversity and the Peopling of the Americas*. Berkeley: University of California Press, 1995.

———. "Linguistic Diversity and the First Settlement of the New World." *Language* 66, no. 3 (1990): 475–521.

———. *"The First Four Discoveries of America: Linguistic Evidence."* Paper presented at the Annual Meeting for the American Association for the Advancement of Science, Philadelphia, 1998.

Nott, Josiah Clark, and George Robins Gliddon. *Types of Mankind; or, Ethnological Researches, Based on Ancient Monuments, Paintings, Sculptures, and Crania of Races*. 2nd ed. Philadelphia: Lippincott, Grambo, 1854.

———. *Indigenous Races of the Earth*. Philadelphia: Lippincott, 1857.

Obama, Barack. *Inaugural Address*, January 20, 2009. https://obamawhitehouse. archives.gov/blog/2009/01/21/president-barack-obamas-inaugural-address.

Olsen, Bjornar J. "The End of History? Archaeology and the Politics of Identity in a Globalised World." In *Destruction and Conservation of Cultural Property*, edited by Robert Layton, Peter G. Stone, and Julian Thomas, 42–54. London: Routledge, 2001.

Opfer, Chris. "Body of Evidence: Technology Takes the Guesswork Out of Identifying Skeletal Remains." *Popular Science*, July 2010, 34.

Oreskes, Naomi. *The Rejection of Continental Drift: Theory and Method in American Earth Science*. New York: Oxford University Press, 1999.

Ousley, Stephen, Richard Jantz, and Donna Freid. "Understanding Race and Human Variation: Why Forensic Anthropologists are Good at Identifying Race." *American Journal of Physical Anthropology* 139 (2009): 68–76.

Ousley, Stephen D., William T. Billeck, and R. Eric Hollinger. "Federal Repatriation Legislation and the Role of Physical Anthropology in Repatriation." *American Journal of Physical Anthropology* 128, S41 (2005): 2–32.

Owen, R. "The Americas: The Case Against an Ice-Age Human Population." In *The Origins of Modern Humans: A World Survey of the Fossil Evidence*, edited by Fred H. Smith and Frank Spencer, 517–63. New York: R. Liss, 1984.

Owsley, Douglas W. "From Jamestown to Kennewick: An Analogy Based on Early Americans." In *Who Were the First Americans: Proceedings of the 58th Annual Biology Colloquium*, edited by Robson Bonnichsen, 128. Corvallis: Center for the Study of the First Americans, Oregon State University.

Owsley, Douglas W., and Richard L. Jantz. "Archaeological Politics and Public Interest in Paleoamerican Studies: Lessons from Gordon Creek Woman and Kennewick Man." *American Antiquity* 66, no. 4 (October 2001): 565–75.

———. "Kennewick Man—A Kin Too Distant?" In *Claiming the Stones/Naming the Bones: Cultural Property and the Negotiation of National and Ethnic Identity*, edited by Elazar Barkan and Ronald Bush, 141–61. Los Angeles: Getty Research Institute, 2002.

Paabo, S. "Molecular Genetic Methods in Archaeology: A Prospect." *Anthropologischer Anzeiger* 45, no. 1 (1987): 9–17.

Paabo, S., R. G. Higuchi, and A. C. Wilson. "Ancient DNA and the Polymerase Chain Reaction: The Emerging Field of Molecular Archaeology." *The Journal of Biological Chemistry* 264, no. 17 (1989): 9709–12.

Paracelsus, quoted in *Memoirs Read Before the Anthropological Society of London, 1863–64*, vol. 1, edited by Anthropological Society of London, 354. London: Trubner and Co, 1865.

Pensley, D. S. "The Native American Graves Protection and Repatriation Act (1990): Where the Native Voice Is Missing." Twentieth Anniversary Commemorative Issue, *Wicazo Sa Review* 20, no. 2 (Autumn 2005): 37–64.

Peregrine, Peter Neal. *Attachment to* November 23, 2010, email. Accessed December 13, 2010. http://www.unl.edu/rhames/AAA/AAA-LRP.pdf.

Pickering, Andrew. *The Mangle of Practice: Time, Agency, and Science*. Chicago: The University of Chicago Press, 1995.

Pijoan, Terese. *American Indian Creation Myths. Santa Fe*, NM: Sunstone Press, 2005.

Popper, Karl R. *The Logic of Scientific Discovery*. London: Routledge, 1992. First published in English in 1959 by Hutchinson and Company.

Powell, Joseph F. "Dental Evidence for the Peopling of the New World: Some Methodological Considerations." *Human Biology* 65 (1993): 799–815.

———. *The First Americans: Race, Evolution, and the Origin of Native Americans*. Cambridge, UK: Cambridge University Press, 2005.

Powell, Joseph F., and D. Gentry Steele. "A Multivariate Craniometric Analysis of North American Paleoindian Remains." *Current Research in the Pleistocene* 9 (1992): 59–62.

Preston, Douglas. "The Mystery of Sandia Cave." *The New Yorker*, June 12, 1995, 66–83.

Priest, Josiah. *American Antiquity and Discoveries in the West. Albany, NY: Hoffman and White*, 1835.

Pringle, Heather. "Battle for the Xinjiang Mummies." *Archaeology* 63, no. 4 (July/August 2010): 30–35.

Prucha, Francis R. *The Great Father: The United States Government and the American Indians. Abridged ed*. Lincoln: University of Nebraska Press, 1986.

Public Broadcasting Service. *"Meet Kennewick Man."* http://www.pbs.org/wgbh/
nova/first/kennewick2.html.

Raff, Jennifer. *"What the Ancient DNA Discovery Tells Us About Native American
Ancestry."* *Guardian*, January 3, 2018. https://www.theguardian.com/science/2018/
jan/03/what-the-ancient-dna-discovery-tells-us-about-native-american-ancestry.

Rafinesque, Constantine Samuel. *The American Nations; or, Outlines of Their General
History, Ancient and Modern: Including the Whole History of the Earth and Mankind in the Western Hemisphere; the Philosophy of American History; the Annals,
Traditions, Civilizations, Languages, etc., of All the American Nations, Tribes, Em-
pires, and States. 2 vols.* Philadelphia: Author, 1836.

Ramenofsky, A. "Developments in the Problem of the North American Paleo-Indian."
*Smithsonian Miscellaneous Collections* 100 (1940): 51–116.

Rassmussen, Knud. *The Eagle's Gift: Alaska Eskimo Tales, translated by Isobel
Hutchinson. Doubleday, Doran & Company*, 1932.

Ray N., D. Wegmann, N. J. R. Fagundes, S. Wang, A. Ruiz-Linares, and L. Excoffier.
"A Statistical Evaluation of Models for the Initial Settlement of the American Con-
tinent Emphasizes the Importance of Gene Flow with Asia." *Molecular Biology and
Evolution* 27 (2010): 337–45.

Reid, Lynette. "Networking Genetics, Populations, and Race." *The American Journal
of Bioethics* 9, no. 6–7 (2009): 50.

Renfrew, Colin. *Archaeology and Language. London: Jonathan Cape*, 1987.

Renfrew, Colin, April McMahon, and Larry Trask, eds. Time *Depth in Historical Linguistics.* Cambridge, UK: McDonald Institute for Archaeological Research, 2000.

Rice, Jocelyn. "20 Things You Didn't Know about Viruses." *Discover* 31, no. 3
(2010): 80.

Richards, M., B. C. Sykes, and R. E. M. Hedges. "Authenticating DNA Extracted
from Ancient Skeletal Remains." *Journal of Archaeological Science* 22 (1995): 291–99.

Richman, Jennifer R. *"NAGPRA: Constitutionally Adequate?"* In *Legal Perspectives on
Cultural Resources*, edited by Jennifer R. Richman and Marion P. Forsyth, 216–31.
Walnut Creek, CA: AltaMira, 2004.

Riding In, James. "Without Ethics and Morality: A Historical Overview of Imperial Archaeology and American Indians." *Arizona State Law Journal* 24, no. 1 (1992): 11–34.

Robbins, Jeffery, ed. *The Pleasure of Finding Things Out: The Best Short Works of Richard Feynman.* New York: Basic Books, 1999.

Roberts, Alexa. "Trust Me, I Work for the Government: Confidentiality and Public
Access to Sensitive Information." *American Indian Quarterly* 25, no. 1 (Spring
2001): 13–17.

Roberts, Frank H. H., Jr. "Developments in the Problem of the North American
Paleo-Indian." In "Essays in Historical Anthropology of North America." *Smithsonian Miscellaneous Collections* 100 (1940): 51–116.

———. "Evidence for a Paleo-Indian in the New World." *Acta Americana I*, no. 2
(1943): 171–201.

———. "Radiocarbon Dates and Early Man." In *Radiocarbon Dating: A Report on the Program to Aid in the Development of the Method of Dating*, edited by F. Johnson, 20–23. Salt Lake City: Society for American Archaeology, 1951.

———. "*Recent Evidence Relating to an Early Indian Occupation of North America.*" *Proceedings of the Eighth American Scientific Congress held in Washington May 10–18, 1940*, edited by Paul H. Oehser, 31–38. Washington, DC: Department of State, 1941–43.

Rogers, R. A., and L. Martin. "The Folsom Discovery and the Concept of Breakthrough Sites in Paleoindian Studies." *Current Research in the Pleistocene* 4 (1987): 81–82.

Rogers, R. A., L. A. Rogers, R. S. Hoffman, and L. D. Martin. "Native American Biological Diversity and the Biogeographic Influence of Ice Age Refugia." *Journal of Biogeography* 18 (1991): 623–30.

Rogers, R. A., L. A. Rogers, and L. D. Martin. "How the Door Opened: The Peopling of the New World." *Human Biology* 64 (1992): 281–302.

Rogers, Richard A. "Glacial Geography and Native North American Languages." *Quaternary Research* 23 (1985): 130–37.

———. "Language, Human Subspeciation, and Ice Age Barriers in Northern Siberia." *Canadian Journal of Anthropology* 5 (1986): 11–22.

———. "Wisconsinan Glaciation and the Dispersal of Native Ethnic Groups in North America." In *Woman, Poet, Scientist: Essays in New World Anthropology Honoring Dr. Emma Lou Davis*, edited by Thomas C. Blackburn, 105–13. San Diego: Ballena Press, 1985.

Romelt, Alfred S. "Pleistocene Vertebrates and Their Bearing on the Problem of Human Antiquity in North America." In *The American Aborigines: Their Origin and Antiquity*, edited by Diamond Jenness, 49–83. Toronto: University of Toronto Press, 1933.

Roosevelt, A. C., M. Lima da Costa, C. Lopes Machado, M. Michab, N. Mercier, H. Valladas, J. Feathers, et al. "Paleoindian Cave Dwellers in the Amazon: The Peopling of the Americas." *Science* 272 (1996): 373–84.

Rose, Jerome C., Thomas J. Green, and Victoria D. Green. "Osteology and the Repatriation of Skeletons." *Annual Review of Anthropology* 25 (1996): 81–103.

Roseman, Charles C., and Timothy D. Weaver. "Multivariate Apportionment of Global Human Craniometric Diversity." *American Journal of Physical Anthropology* 125 (2004): 257–63.

Rothenberg, Marc, ed. "*Joseph Henry: Promoter of Anthropology.*" *Smithsonian Institution Archives.* Accessed February 17, 2011. http://siarchives.si.edu/history/jhp/joseph08.htm.

Rothstein, Edward. "Another Stop on a Long, Improbable Journey." *New York Times*, February 20, 2011.

Rowlands, Michael. "The Politics of Identity in Archaeology." In *Social Construction of the Past: Representation as Power*, edited by George Bond and Angela Gilliam, 129–43. London: Routledge, 1994.

Rudolph, T., ed. *The Hetrick Site: 11,000 Years of Prehistory in the Weiser River Valley, Idaho*. Report on file at the Idaho Historical Society, Boise, 1995.

"*Russian Money Feeds International Art Boom.*" *Artinfo*. Accessed May 6, 2011. http://www.artinfo.com/news/story/25136/russian-money-feeds-international-art-boom/.

Safier, Neil. "Global Knowledge on the Move: Itineraries, Amerindian Narratives, and Deep Histories of Science." *Isis*, 1001 (2010): 133–45.

Said, Edward. *Culture and Imperialism*. New York: Alfred A. Knopf, 1993.

———. "Representing the Colonized: Anthropology's Interlocutors." *Critical Inquiry* 15 (1989): 205–25.

Sauer, Carl O. "A Geographic Sketch of Early Man in America." *Geographical Review* 34, no. 4 (October 1944): 529–73.

Sauer, Norman J. "Forensic Anthropology and the Concept of Race: If Races Don't Exist, Why are Forensic Anthropologists So Good at Identifying Them?" *Social Science and Medicine* 34, no. 2 (1992): 107–11.

Schorr, T. G., S. W. Ballinger, Y. Y. Gan, J. A. Hodge, D. A. Merriwether, D. N. Lawrence, W. C. Knowler, K. M. Weiss, and D. C. Wallace. "Amerindian Mitochondrial DNAs Have Rare Asian Variants at High Frequencies, Suggesting They Derived from Four Primary Maternal Lineages." *American Journal of Human Genetics* 46 (1990): 613–23.

Schroeder, K. B. "A Private Allele Ubiquitous in the Americas." *Biology Letters* 3/2 (2007): 218–23.

Schroeder, K. B., M. Jakobsson, M. H. Crawford, T. G. Schurr, S. M. Boca, D. F. Conrad, R. Y. Tito, et al. "Haplotypic Background of a Private Allele at High Frequency in the Americas." *Molecular Biology Evolution* 26 (2009): 995–1016.

Schurr, T. G., and S. T. Sherry. "Mitochondrial DNA and Y Chromosome Diversity and the Peopling of the Americas: Evolutionary and Demographic Evidence." *American Journal of Human Biology* 16 (2004): 420–39.

Schweger, C., J. Matthews, D. Hopkins, and S. Young. "Paleoecology of Beringia—A Synthesis." In *Paleoecology of Beringia*, edited by David M. Hopkins, John V. Matthews, Charles E. Schweger, and Steven B. Young, 425–44. New York: Academic Press, 1982.

Semali, Ladislaus, and Joe Kincheloe. "*Introduction: What Is Indigenous Knowledge and Why Should We Study It?*" In *What Is Indigenous Knowledge? Voices from the Academy*, edited by Ladislaus M. Semali and Joe L. Kincheloe, 3–57. New York: Fainter Press, 1999.

Shapin, Steven, and Simon Schaffer. *Leviathan and the Air-Pump: Hobbes, Boyle, and the Experimental Life*. Princeton, NJ: Princeton University Press, 1985.

Shields, G. F., A. M. Schmiechen, B. L. Frazier, A. Redd, M. I. Voevoda, J. K. Reed, and R. H. Ward. "mtDNA Sequences Suggest a Recent Evolutionary Divergence for Beringian and Northern North American Populations." *American Journal of Human Genetics* 53 (1993): 549–62.

Shott, Michael. "Crises and Solutions in American Archaeology." *SAA Archaeology Record* 10, no. 1 (January 2010): 37–38.

Silverberg, Robert. *The Mound Builders*. Athens: Ohio University Press, 1970.

Smith, D. G., R. S. Malhi, J. A. Eshleman, F. A. Kaestle, and B. M. Kemp. "Mito-chondrial DNA Haplogroups of Paleoamericans in North America." *Paleoamerican Origins: Beyond Clovis*, edited by Robson Bonnichsen, Brad Lepper, Dennis Stanford, and Michael Waters, 243–54. College Station: Texas A&M University Press, 2005.

Smith, D. G., R. S. Malhi, J. A. Eshleman, J. G. Lorenz, and F. A. Kaestle. "Distribu-tion of mtDNA Haplogroup X among Native North Americans." *American Journal of Physical Anthropology* 110, no. 3 (1999): 271–84.

Smithsonian Institution. "*Visitor Stats*." Accessed September 2, 2019. https://www.si.edu/newsdesk/about/stats.

———. http://anthropolgy.si.edu/outreach/indbibl/americanindianlanguages.pdf.

Smithsonian Institution National Museum of the American Indian, Office of Public Affairs. "*Listening to Our Ancestors*." http://americanindian.si.edu/press/releases/2006-01-23_listening_to_our_ancestors_press_release_for_web.pdf.

Society for American Archaeology (SAA). "*2009 Repatriation Issues*." http://www.saa.org/AbouttheSociety/GovernmentAffairs/RepatriationIssuesitabid/214/Default.aspx.

———. "Society for American Archaeology, Principles of Archaeological Ethics." *American Antiquity* 61, no. 3 (1996): 451–52.

Squier, E. G., and E. H. Davis. *Ancient Monuments of the Mississippi Valley: Compris-ing the Results of Extensive Original Surveys and Explorations. Washington, DC: Smithsonian Institution*, 1848.

Stanford, D. "Clovis Origins and Adaptations: An Introductory Perspective." In *The First Americans: Search and Research*, edited by Thomas D. Dillehay and David J. Meltzer, 1–13. Boca Raton, FL: CRC Press, 1991.

———. "Pre-Clovis Occupation South of the Ice Sheets." In *Early Man in the New World*, edited by Richard Shutler, 65–72. Beverly Hills, CA: Sage, 1983.

Stanford, Dennis, and Bruce Bradley. *Across Atlantic Ice: The Origin of America's Clo-vis Culture*. Berkeley: University of California Press, 2012.

———. "Ocean Trails and Prairie Paths? Thoughts about Clovis Origins." In *The First Americans: The Pleistocene Colonization of the New World*, edited by Nina G. Jablonski, 255–72. San Francisco: California Academy of Sciences, 2002.

Stanton, William R. *The Leopard's Spots: Scientific Attitudes toward Race in America, 1815–1859*. Chicago: University of Chicago Press, 1960.

Stapp, Darby C. "The Impact of NAGPRA on Anthropology: A View from Cultural Resource Management in the Pacific Northwest." In *Opening Archaeology: Repatri-ation's Impact on Contemporary Research and Practice*, edited by Thomas W. Killion, 212. Santa Fe, NM: School for Advanced Research Press.

Stapp, Darby C., and Michael S. Burney. *Tribal Cultural Resource Management: The Full Circle to Stewardship*. Walnut Creek, CA: AltaMira, 2002.

Starikovskaya, Y. B., R. I. Sukernik, T. G. Schurr, and D. C. Wallace. "Mitochon-drial DNA Diversity in Chukchi and Siberian Eskimos: Implications for the

Genetic Prehistory of Ancient Beringia." *American Journal of Human Genetics 63* (1998): 1473–91.

Steele, D. Gentry, and Joseph F. Powell. "Paleobiology of the First Americans." *Evolutionary Anthropology* 2 (1993): 138–46.

———. "Peopling of the Americas: Paleobiological Evidence." *Human Biology* 64 (1992): 303–36.

Stewart, T. D. "The Development of the Concept of Morphological Dating in Connection with Early Man in America." *Southwestern Journal of Anthropology* 5, no. 1 (1949): 1–16.

———. "The Evolutionary Status of the First Americans." *American Journal of Physical Anthropology* 56 (1981): 461–66.

———. "A Physical Anthropologist's View of the Peopling of the New World." *Southwest Journal of Anthropology* 16 (1960): 259–73.

Stone, A. C., and M. Stoneking. "Genetic Analyses of an 8000-Year-Old Native American Skeleton." *Ancient Biomolecules* 1, no. 2 (1996): 83–87.

———. "mtDNA Analysis of a Prehistoric Oneota Population: Implications for the Peopling of the New World." *American Journal of Human Genetics* 62 (1998): 1153–70.

Strassberg, Richard E., ed. and trans. *A Chinese Bestiary: Strange Creatures from the Guideways through Mountains and Seas*. Berkeley: University of California Press, 2002.

Strauss, Lawrence Guy. "Solutrean Settlement of North America? A Review of Reality." *American Antiquity* 65 (2000): 219.

Strauss, Lawrence Guy, David J. Meltzer, and Ted Goebel. "Ice Age Atlantis? Exploring the Solutrean-Clovis 'Connection.'" *World Archaeology* 37 (2005): 506–31.

Swidler, Nina, Kurt E. Dongoske, Roger Anyon, and Alan Downer, eds. *Native Americans and Archaeologists: Stepping Stones to Common Ground*. Walnut Creek, CA: AltaMira, 1997.

Szathmary, E. J. E. *"Ancient Migrations from Asia to North America."* *Prehistoric Mongoloid Dispersals*, edited by Takeru Akazawa and Emoke J. E. Szathmary, 149–64. Oxford, UK: Oxford University Press, 1996.

———. "Genetics of Aboriginal North Americans." *Evolutionary Anthropology* 1 (1993): 202–20.

———. "Modelling Ancient Population Relationships from Modern Population Genetics." In *Method and Theory for Investigating the Peopling of the Americas*, edited by R. Bonnichsen and D. Gentry Steele, 117–30. Corvallis: Center for the Study of the First Americans, Oregon State University, 1994.

———. "Peopling of Northern North America: Clues from Genetic Studies." *Acta Anthropologia* 8 (1984): 79–109.

———. "Peopling of North America: Clues from Genetic Studies." In *Out of Asia: Peopling the Americas and the Pacific*, edited by R. Kirk and E. Szathmary, 79–104. Canberra: The Journal of Pacific History, 1985.

Tankersley, Kenneth. In *Search of Ice Age Americans*. Salt Lake City: Gibbs Smith Publisher, 2002.

Tarazona-Santos, E., and F. R. Santos. "The Peopling of the Americas: A Second Major Migration?" *American Journal of Human Genetics* 70 (2002): 1377–80.

Taylor, R., L. Payen, C. Prior, P. Slota, R. Gillespie, J. Gowlett, R. Hedges, et al. "Major Revisions in the Pleistocene Age Assignments for North American Human Skeletons by C-14 Accelerator Mass Spectrometry: None Older Than 11,000 C-14 Years B.P." *American Antiquity* 50 (1985): 136–40.

Thomas, David Hurst. "American Archaeology in the Twenty-First Century." In *Opening Archaeology: Repatriation's Impact on Contemporary Research and Practice*, edited by Thomas W. Killion, 58. Santa Fe, NM: School for Advanced Research Press, 2008.

———. *Skull Wars: Kennewick, Archaeology, and the Battle for Native American Identity*. New York: Basic Books, 2000.

———. "Thomas Jefferson's Conflicted Legacy in American Archaeology." In *Across the Continent: Jefferson, Lewis and Clark, and the Making of America*, edited by Douglas Seefeldt, Peter Onuf, and Jeffrey Hartman, 84–131. Charlottesville: University of Virginia Press, 2005.

Thomason, Sarah, and Terrence Kaufman. *Language Contact, Creolization, and Genetic Linguistics*. Berkeley: University of California Press, 1988.

Thompson, J. Eric S. "The First Peopling of the New World." *American Antiquity* 24, no. 2 (1958): 184.

Tierney, John. "Social Scientist Sees Bias Within." *New York Times*, February 8, 2011. http://www.nytimes.com/2011/02/08/science/08tier.html?ref=science.

Todd, Brenda Kaye. "*The Disconnection between Anthropological Theories of Ethnicity and Identity and the Definition of 'Cultural Affiliation' under NAGPRA*." Master's thesis, University of Colorado, Boulder, 2005.

Torroni, A., J. V. Neel, R. Barrantes, T. G. Schurr, and D. C. Wallace. "A Mitochondrial DNA 'Clock' for the Amerinds and Its Implications for Timing Their Entry into North America." *Proceedings of the National Academy of Sciences* 91 (1994): 1158–62.

Torroni, A., R. I. Sukernik, T. G. Schurr, Y. B. Starikovskaya, M. A. Cabell, M. H. Crawford, A. G. Comuzzie, and D. C. Wallace. "mtDNA Variation of Aboriginal Siberians Reveals Distinct Genetic Affinities with Native Americans." *American Journal of Human Genetics* 53 (1993): 591–608.

Torroni, A., T. G. Schurr, M. F. Cabell, M. D. Brown, J. V Neel, M. Larsen, D. G. Smith, C. M. Vullo, and D. C. Wallace. "Asian Affinities and Continental Radiation of the Four Founding Native American mtDNAs." *American Journal of Human Genetics* 53 (1993): 563–90.

Torroni, A., T. G. Schurr, C. C. Yang, E. Szathmary, R. C. Williams, M. S. Schanfield, G. A. Troup, et al. "Native American Mitochondrial DNA Analysis Indicates that the Amerindian and the Nadene Populations Were Founded by Two Independent Migrations." *Genetics* 130 (1992): 153–62.

Trigger, Bruce G. "Archaeology and the Image of the American Indian." *American Antiquity* 45, no. 4 (1980): 662–76.

———. "Prehistoric Archaeology and American Society: An Historical Perspective." In *American Archaeology: Past and Future*, edited by D. Meltzer, D. D. Fowler, and A. Sabloff, 187–215. Washington, DC: Smithsonian Institution Press, 1986.

Trope, J. F., and W. R. Echo-Hawk. "The Native American Graves Protection and Repatriation Act: Background and Legislative History." *Arizona State Law Journal* 24, no. 1 (1992): 35–78.

Turner, C. G. "Dental Evidence for the Peopling of the Americas." In *Early Man in the New World*, edited by Richard Shutler, 147–57. Beverly Hills, CA: Sage Publications, 1983.

———. "Dental Evidence for the Peopling of the Americas." *National Geographic Society Research Reports* 19 (1985): 573–96.

———. "The Dental Search for Native American Origins." In *Out of Asia: Peopling the Americas and the Pacific*, edited by Robert Kirk and Emoke Szathmary, 31–78. Canberra: Australian National University, The Journal of Pacific History,1985.

———. "Dentochronological Separation Estimates for Pacific Rim Populations." *Science* 23 (1986): 1140–42.

———. "The First Americans: The Dental Evidence." *National Geographic Research* 2 (1986): 37–46.

———. "Late Pleistocene and Holocene Population History of East Asia Based on Dental Variation." *American Journal of Physical Anthropology* 73 (1987): 305–21.

———. "Major Features of Sundadonty and Sinodonty, Including Suggestions about East Asian Microevolution, Population History, and Late Pleistocene Relationships with Australian Aboriginals." *American Journal of Physical Anthropology* 82 (1990): 295–317.

———. "The Modern Human Dispersal Event: The Eastern Frontier." *The Quarterly Review of Archaeology* 6 (1985): 8–9, 13.

———. "Sinodonty and Sundadonty: A Dental Anthropological View of Mongoloid Microevolution, Origin, and Dispersal into the Pacific Basin, Siberia, and the Americas." In *Late Pleistocene and Early Holocene Cultural Connections of Asia and America*, edited by R. S. Vasilievsky, 72–76. Siberian Branch, Novosibirsk: USSR Academy of Sciences, 1983.

———. "Telltale Teeth." *Natural History* 96, no. 1 (1987): 6–10.

———. "Three-Rooted Mandibular First Permanent Molars and the Question of American Indian Origins." *American Journal of Physical Anthropology* 34 (1971): 229–41.

Turner, C. G., and J. Bird. "Dentition of Chilean Paleo-Indians and the Peopling of the Americas." *Science* 212 (1981): 1053–54.

Tuross, Noreen, and Tom D. Dillehay. "The Mechanism of Organic Preservation at Monte Verde, Chile, and One Use of Miomolecules in Archaeological Interpretation." *Journal of Field Archaeology* 22, no. 1 (Spring 1995): 97–110.

Twain, Mark (Samuel Clemens). *A Connecticut Yankee in King Arthur's Court*, 468. https://www.google.com/books/edition/A_Connecticut_Yankee_in_King_Arthur_s_Co/vQvlogplcaYC?hl=en.

United States Capitol Visitor Center. https://www.aoc.gov/explore-capitol-campus/buildings-grounds/capitol-building/capitol-visitors-center.

United States Department of the Army. *Cultural Resources Management: Environmental Quality. Army Regulation 200-4.* Washington, DC: Department of Army, 1998.

United States Department of the Interior. *Federal Agencies Task Force Report, American Indian Religious Freedom Act Report.* Washington, DC: US Government Printing Office, 1979.

United States House of Representatives. "Providing for the Protection of Native American Graves, and For Other Purposes." *House Report 101-877, to accompany H. R. 5237,* October 15, 1990.

———. *Providing for the Protection of Native American Graves, and For Other Purposes.* 101st Cong., 2d sess., 1990. H. Doc. 101–877.

———. "*Report 4027.*" Accessed January 26, 2021. http://rla.unc.edu/saa/repat/Legislative/hr4027_text.pdf.

United States National Institutes of Health. "*A Nomenclature System for the Tree of Human Y-Chromosomal Binary Haplogroups.*" *US National Library of Medicine.* Accessed July 5, 2019. ncbi.nlm.nih.gov/pmc/articles/PMC155271/#!po=1.92308.

United States National Oceanic and Atmospheric Administration (NOAA). Accessed June 26, 2019. www.ncdc.noaa.gov/.

———. "*Summary of 100,000 Years.*" Accessed January 26, 2021. http://www.ncdc.noaa.gov/paleo/ctl/100k.html.

United States Senate. *Providing for the Protection of Native American Graves and the Repatriation of Native American Remains and Cultural Patrimony. 101st Congress, 2d sess.,* 1990. S. Doc. 101–473.

———. *S. R. 2843.* http://www.gpo.gov/fdsys/pkg/BILLS-108s2843is/pdf/BILLS-108s2843is.pdf.

University of Illinois. "*World Haplogroups Maps.*" Accessed July 5, 2019. http://www.scs.illinois.edu/~mcdonald/WorldHaplogroupsMaps.pdf.

University of Tennessee, Knoxville. "*FORDISC 3.1.*" Accessed July 3, 2019. fac.utk.edu/fordisc-3-1-personal-computer-forensic-discriminant-functions/.

Untermeyer, Louis. *The Poems of Henry Wadsworth Longfellow.* New York: The Heritage Press, 1943.

"US Tribe Stops Study of Bones that Challenge History." *New York Times,* September 30, 1996. http://www.nytimes.com/1996/09/30/us/tribe-stops-study-of-bones-that-challenge-history.html.

Vanderbilt University Department of Anthropology. "*Monte Verde.*" http://www.vanderbilt.edu/AnS/Anthro/Anth210/monte_verde.htm.

Villa, Dana. *Arendt and Heidegger: The Fate of the Political.* Princeton: Princeton University Press, 1996.

Vincent, Joan. *Anthropology and Politics: Visions, Traditions, and Trends*. Tucson: University of Arizona Press, 1990.

Volman, K. C. "Paleoenvironmental Implications of Botanical Data from Meadowcroft Rockshelter, Pennsylvania." *PhD diss., Graduate College of Texas A&M University*, 1981.

Von Franz, Marie-Louise. *Creation Myths*. Boston: Shambhala Publications, 1972.

Wade, Nicholas. "Anthropology as Science: Statement Deepens a Rift." *New York Times*, December 9, 2010.

Wallace, Douglas C., Katherin Garrison, and William C. Knowler. "Dramatic Founder Effects in Amerindian Mitochondria DNAs." *American Journal of Physical Anthropology* 68 (1985): 149–55.

Wallace, D. C., and A. Torroni. "American Indian Prehistory as Written in the Mitochondrial DNA: A Review." *Human Biology* 64, no. 3 (1992): 403–16.

Ward, R. H., A. Redd, D. Valencia, B. Frazier, and S. Paabo. "Genetic and Linguistic Differentiation in the Americas." *Proceedings of the National Academy of Sciences* 90 (1993): 10663–67.

Waters, Michael R. "Early Man in the New World: An Evaluation of the Radiocarbon Dated Pre-Clovis Sites in the Americas." In *Environments and Extinctions*, edited by Jim I. Mead and David J. Meltzer, 125–43. Orono: Center for the Study of Early Man, University of Maine, 1985.

Waters, Michael R., and Thomas W. Stafford Jr. "Redefining the Age of Clovis: Implications for the Peopling of the Americas." *Science* 315 (2007): 1122–26.

Watkins, Joe. *Indigenous Archaeology: American Indian Values and Scientific Practice*. Walnut Creek, CA: AltaMira, 2000.

———. "The Repatriation Arena: Control, Conflict, and Compromise." In *Opening Archaeology: Repatriation's Impact on Contemporary Research and Practice*, edited by Thomas W. Killion, 162. Santa Fe, NM: School for Advanced Research Press, 2008.

Weaver, Jace. "Indian Presence with No Indians Present: NAGPRA and Its Discontents." *Wicazo Sa Review* 12, no. 2 (Autumn 1997): 13–30.

Wendorf, Fred. "The Antiquity of Man in America." In *Late Quaternary Environments of the United States: The Late Pleistocene*, edited by Stephen C. Porter, 364–82. Minneapolis: University of Minnesota Press, 1983.

———. "Early Man in the New World: Problems of Migration." *The American Naturalist* 100 (1966): 253–70.

White, James M., Rolf W. Mathewes, and W. H. Mathews. "Late Pleistocene Chronology and Environment of the 'Ice-Free Corridor' of Northwestern Alberta." *Quaternary Research* 24 (1985): 173–86.

White, Lynn. "The Historical Roots of our Ecological Crisis." *Science* 155, no. 3767 (March 19, 1967): 1203–7.

Whitley, David S., and Ronald I. Dorn. "New Perspectives on the Clovis vs. Pre-Clovis Controversy." *American Antiquity* 58, no. 4 (October 1993): 626–47.

Wilford, John Noble. "On Crete, New Evidence of Very Ancient Mariners." *New York Times*, February 15, 2010. http://www.nytimes.com/2010/02/16/science/16archeo.html.

———. "'American' Arrowhead Found in Siberia." *New York Times*, August 2, 1996.

Willey, Gordon R. "Archeological Theories and Interpretation: New World." In *Anthropology Today*, edited by Sol Tax, 361–85. Chicago: University of Chicago Press, 1953.

Williams, Robert A., Jr. *The American Indian in Western Legal Thought: The Discourses of Conquest*. New York: Oxford University Press, 1990.

Williams, Robert C., Arrthur G. Steinberg, Henry Gershowitz, Peter H., Bennett, William C. Knowler, David J. Pettitt, William Butler, et al. "GM Allotypes in Native Americans: Evidence for Three Distinct Migrations Across the Bering Land Bridge." *American Journal of Physical Anthropology* 66 (1985): 1–19.

Wilmsen, Edwin N. "An Outline of Early Man Studies in the United States." *American Antiquity* 31, no. 2 (1965): 172–92.

Wilson, Thomas. "Address: The Beginnings of the Science of Prehistoric Anthropology." *Proceedings of the American Association for the Advancement of Science* 48 (1899): 307–53.

———. "Results of an Inquiry as to the Existence of Man in North America during the Paleolithic Period of the Stone Age." In *Annual Report for the Year Ending June 30, 1888, United States National Museum, 677–702. Washington, DC: Smithsonian Institution, United States National Museum*, 1890.

———. "Report on the Department of Prehistoric Anthropology in the U.S. National Museum." In *Annual Report of the United States National Museum for 1888*, 123–38. Washington, DC: Smithsonian Institution, United States National Museum, 1890.

———. "A Study of Prehistoric Anthropology: Handbook for Beginners." In *Annual Report for the Year Ending June 30, 1888, United States National Museum, 597–671. Washington, DC: Smithsonian Institution, United States National Museum*, 1890.

Winsor, Justin. "Aboriginal America." In *Narrative and Critical History of North America*, vol. 1, edited by Justin Winsor. Cambridge, MA: Houghton Mifflin, 1889.

———. "The Progress of Opinion Respecting the Origin and Antiquity of Man in America." In *Narrative and Critical History of America*, vol.1, edited by Justin Winsor, 369–412. Cambridge, MA: Houghton Mifflin Company, 1889.

Winsor, Justin, ed. *Ancient Man in North America*. Denver: Colorado Museum of Natural History, 1939.

Wright, George Frederick. "Mr. Holmes's Criticism upon the Evidence of Glacial Man." *Science* 21, no. 537 (1893): 267–68.

———. *Man and the Glacial Period. New York: Appleton*, 1892.

Wright, H. E., Jr., J. E. Kutzbach, T. Webb III, W. F. Ruddiman, F. A. Street-Perrott, and P. J. Bartlein. *Global Climates since the Last Glacial Maximum*. Minneapolis: University of Minnesota Press, 1993.

Zegura, S. "The Initial Peopling of the Americas: An Overview." In *Out of Asia: Peopling the Americas and the Pacific*, edited by R. Kirk and E. Szathmary, 1–18. Canberra: The Journal of Pacific History, 1985.

Zegura S. L., T. M. Karafet, L. A. Zhivotovsky, and M. F. Hammer. "High-Resolution SNPs and Microsatellite Haplotypes Point to a Single, Recent Entry of Native American Y Chromosomes into the Americas." *Molecular Biology and Evolution* 21 (2004): 164–75.

Zimmerman, Larry L. "Multivocality, Descendant Communities, and Some Epistemological Shifts Forced by Repatriation." In *Opening Archaeology: Repatriation's Impact on Contemporary Research and Practice*, edited by Thomas W. Killion, 97. Santa Fe: School for Advanced Research Press, 2008.

Zorich, Zach. "Undiscovery of the Year: Clovis Comet." Archaeology 64, no. 1 (January/February 2011).

# INDEX

ROBERT V. DAVIS JR. holds a PhD in science and technology studies from Virginia Tech University. Since his retirement from public service with the federal government, he has focused his research and writing on the interactions between science and public policy.

CPSIA information can be obtained
at www.ICGtesting.com
Printed in the USA
LVHW030259231221
707027LV00001B/63